Biotechnology and its Applications in AGRICULTURAL SCIENCE

Editor
P.B. Bansal

2025

GENE-TECH BOOKS

ISBN 978-81-89729-05-9

Published by: **GENE-TECH BOOKS**®
4762-63/23, Ansari Road, Darya Ganj,
New Delhi - 110 002
Phone: 09899295259, 011-43003222
E-mail: genetechbooks@yahoo.co.in

PRINTED IN INDIA

Biotechnology and its Applications in AGRICULTURAL SCIENCE

which provides information on biosafety methods and procedures, and on the status of the regulatory frameworks in individual countries.

OECD has established a network of international organizations with activities related to biosafety to enhance the exchange of information and to facilitate cooperation.

IBS provides a number of services to enhance developing country capacities to manage agricultural biotechnology. It gives short training courses annually for managers of agricultural biotechnology research programs in Asia. Biosafety is a key element in these courses. IBS also provides advice to individual countries.

USAID is involved in a number of biosafety training activities in Asian countries, some of which have been provided by its Agricultural Biotechnology Services Program.

ISAAA is currently setting up a Global Knowledge Center on Crop Biotechnology based at ISAAA's SEAsia Center at Los Baños in the Philippines. ISAAA also organizes capacity-building workshops and training for regulatory officials and scientists from the Asian region.

Genetically Modified Organisms

In industrialized countries, considerable progress has been made in methods, approaches, and experience in the safe management of GMOs. Overall these methods have proven to be effective, particularly with respect to food safety. The development of a transgenic soybean containing a protein derived from Brazil nuts, potentially useful in animal feed, was abandoned when the safety assessment revealed that the protein was probably an allergen. In response to consumer concerns, antibiotic markers are now being phased out as alternatives are being developed. That despite consensus among nations

Preface

Biotechnology can be broadly defined as 'using organisms or their products for commercial purposes'. As such, biotechnology has been practices since he beginning of records history. It has been used to bake bread, brew alcoholic beverages, and breed food crops or domestic animals. But recent developments in molecular biology have given biotechnology new meaning, new prominence, and new potential. Modern biotechnology is having a dramatic effect on the world economy and society.

The recent development in biotechnology have led to rapid progress in understanding the genetic basis of living organisms, and the ability to develop products and processes useful to human and animal health, food and agriculture, and industry. In agriculture, there is increasing use of biotechnology for genetic mapping and marker-assisted selection to aid more precise and rapid development of new strains of improved crops and livestock. Biotechnology applications such as tissue culture and micro-propagation are being used for the rapid multiplication of disease-free planting materials.

Biotechnology has given us a new tool to improve food security and reduce poverty. It has offered opportunities to produce more nutritious and better tasting foods, higher crop yields and plants that are naturally protected from disease and insects. It allows for the transfer of only one or a few desirable genes, thereby permitting scientists to develop crops with specific beneficial traits and reduce undesirable traits. It also offers effective techniques to address food safety concerns.

However, the development of agricultural biotechnology is perceived by some as posing considerable risks to human health and the environment. Most of the debate on biotechnology has been focused on genetically modified organisms (GMOs). The debates surrounding GMOs has heightened concerns that genetic engineering may in the long run be harmful to human health and the environment unless effective regulatory frameworks are implemented. Indeed, the public and private sectors must manage the introduction and use of biotechnology to maximize benefits and minimize risks.

Designed for students and practitioners of agriculture, genetic engineering and related fields, this book describes new and potential applications of biotechnology in agricultural sector. This unique and up-to-date resource offering students an innovative and valuable presentation of the subject.

Editor

Contents

1

Introduction

Modern science encompasses new developments in the biological, physical, and social sciences. In biology, discoveries over the past 20 years allow the better understanding of the structure and function of human, animal, and plant genes. At the same time, new discoveries in the physical sciences underpin the revolution in information and communications technologies. Geographic information systems enable characterization of agro-ecosystems and offer means by which new technologies can be customized to the needs of particular agro-ecosystems.

The biological and physical sciences also interact in new ways. For example, the ability to analyze large volumes of data is a critical component of various genome projects that are mapping all the genes in an organism, as in the Human Genome Project. New developments in the social sciences underpin community participation in technology development and evaluation (sometimes termed agro-ecological methods).

Participatory methods developed in the social sciences can help in understanding problems and the researchable issues, particularly those of small farmers operating in marginal environments. They may also be used to clarify the concerns of rural and urban dwellers in regard to the deployment of new technologies, including the products of biotechnology.

Integration of all branches of modern science and traditional knowledge is required to develop knowledge-intensive solutions to the problems of rural Asia. These solutions need not be only technically feasible but also socially acceptable. Indeed, the potential value of modern science to agriculture and the environment in Asia will require the efforts of all stakeholders, including civil society, farmer cooperatives, producers, consumers, governments, and development agencies.

History, and Scope of Biotechnology

Biotechnology, broadly defined, includes any technique that uses living organisms, or parts of such organisms, to make or modify products, to improve plants or animals, or to develop microorganisms for specific use. It ranges from traditional biotechnology to the most advanced modern biotechnology. Biotechnology is not a separate science but rather a mix of disciplines (genetics, molecular biology, biochemistry, embryology, and cell biology) converted into productive processes by linking them with such practical disciplines as chemical engineering, information technology, and robotics.

Modern biotechnology should be seen as an integration of new techniques with the well-established approaches of traditional biotechnology such as plant and animal breeding, food production, fermentation products and processes, and production of pharmaceuticals and fertilizers.

The key components of modern biotechnology are listed below.

(i) *Genomics*: The molecular characterization of all genes in a species.

(ii) *Bioinformatics*: The assembly of data from genomic analysis into accessible forms, involving the application of information technology to analyze and manage large data sets resulting from gene sequencing or related techniques.

(iii) *Transformation*: The introduction of one or more genes conferring potentially useful traits into plants, livestock, fish and tree species.

(iv) Genetically improved organism.

(v) Genetically modified organism (GMO).

(vi) Living modified organism (LMO).

(vii) *Molecular breeding*: Identification and evaluation of useful traits in breeding programs by the use of marker-assisted selection (MAS);

(viii) *Diagnostics*: The use of molecular characterization to provide more accurate and quicker identification of pathogens; and

(ix) *Vaccine technology*: The use of modern immunology to develop recombinant deoxyribonucleic acid (rDNA) vaccines for imprved control of livestock and fish diseases.

Biotechnology consists of a gradient of technologies, ranging from the long-established and widely used techniques of traditional biotechnology to novel and continuously evolving modem biotechnology techniques (Figure 1). During the 1970s scientists developed new methods for precise recombination of portions of deoxyribonucleic acid (DNA), the biochemical material in all living cells that governs inherited characteristics, and for transferring portions of DNA from one organism to another. This set of enabling techniques is referred to as rDNA technology or genetic engineering.

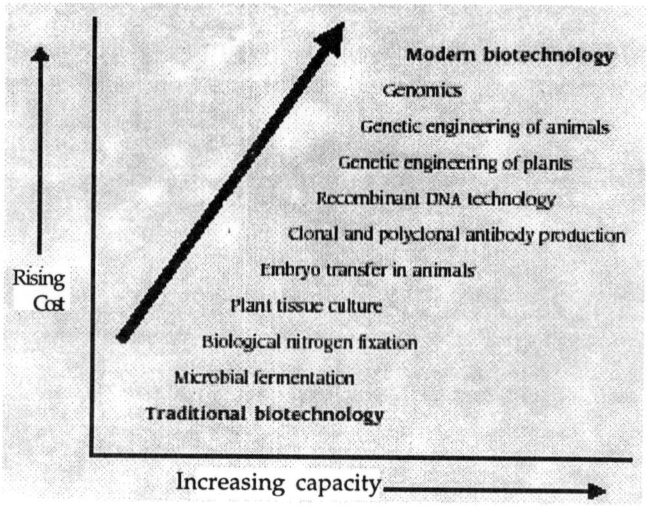

Figure 1. Gradient of Biotechnologies

Modern biotechnology presently includes the various uses of new techniques of rDNA technology, monoclonal and polyclonal antibodies, and new cell and tissue culture methods. A chronology of the development of modern biotechnology is given in Table 1. Over the past two decades the number of significant advances in modern biotechnology for understanding and modifying the genetics of living organisms has increased dramatically. That has led to greatly increased interest and investment in biotechnology, and increasing concerns as to the power of the new technologies and their safety.

Table 1. The Evolution of the Science of Genetics, Leading to Modern Biotechnology

1866	Mendel postulates a set of rules to explain the inheritance of

biological characteristics in living organisms.

1900 Mendelian law rediscovered after independent experimental evidence confirms Mendel's basic principles.

1903 Sutton postulates that genes are located on chromosomes.

1910 Morgan's experiments prove genes are located on chromosomes.

1911 Johannsen devises the term "gene", and distinguishes genotypes (determined by genetic composition) and phenotypes (influenced by environment).

1922 Morgan and colleagues develop gene mapping techniques and prepare gene map of fruit fly chromosomes, ultimately containing over 2000 genes.

1944 Avery, MacLeod and McCarty demonstrated that genes are composed of DNA rather than protein.

1952 Hershey and Chase confirm role of DNA as the basic genetic material.

1953 Watson and Crick discover the double-helix structure of DNA.

1960 Genetic code deciphered.

1971 Cohen and Boyer develop initial techniques for rDNA technology, to allow transfer of genetic material from one organism to another.

1973 First gene (for insulin production) cloned, using rDNA technology.

1974 First expression in bacteria of a gene cloned from a different species.

1976 First new biotechnology firm established to exploit rDNA technology (Genentech in USA).

1980 USA Supreme Court rules that microorganisms can be patented under existing law (Diamond v. Chakrabarty).

1982 First rDNA animal vaccine approved for sale in Europe (colibacillosis). First rDNA pharmaceutical (insulin) approved for sale in USA and UK. First successful transfer of a gene from one animal species to another (a transgenic mouse carrying the gene for rat growth hormone). First transgenic plant produced, using an agrobacterium transformation system.

1983 First successful transfer of a plant gene from one species to another.

1985 US Patent Office extends patent protection to genetically engineered plants.

1986	Transgenic pigs produced carrying the gene for human growth hormone.
1987	First field trials in USA of transgenic plants (tomatoes with a gene for insect resistance). First field trials in USA of genetically engineered microorganism.
1988	US Patent Office extends patent protection to genetically engineered animals. First GMO approved. Human genome mapping project initiated.
1989	Plant genome mapping projects (for cereals and *Arabidopsis*) initiated.
2000	Plant genome mapping projects for rice and Arabidopsis completed, and about 44 million hectares of land planted to GMO crops.

DNA = deoxyribonucleic acid, GMO = genetically modified organism, rDNA = recombinant

DNA, UK = United Kingdom, USA = United States of America.

Modern biotechnology R&D has been conducted in an institutional and economic environment that differs significantly from the development of the earlier Green Revolution technologies. While the latter were essentially the prerogative of public research institutions and philanthropic foundations, the application of modern biotechnology to agriculture is essentially a competitive, commercial endeavor in which powerful private sector interests compete. Similarly, while the Green Revolution technologies were essentially dedicated to the public, the strengthening and extension of IP protection, particularly since the conclusion of the Uruguay Round of trade negotiations has increased the private character of biotechnologies.

Multinational companies in the seed, agricultural chemical, pharmaceutical, and food-processing industries play a major role in biotechnology research. They have invested heavily in in-house research facilities, commissioned research, taken equity positions in new biotechnology firms, and entered into contractual arrangements with public research institutions or universities. As a result of mergers and acquisitions in the past few years, the development of new biotechnology applications in agriculture has become increasingly concentrated in the hands of a decreasing number of companies. The dominant companies that operate within global markets are Aventis, AgrEvo, Dow, DuPont, Monsanto, and Syngenta.

Biotechnology R&D has been concentrated in a limited number of industrialized countries, with the United States (US) in the lead in financial and human resources. A growing number of developing countries have invested in biotechnology R&D, but the amounts are small compared to the sums invested by private companies in the industrial world. While private sector investment in agricultural research in general is increasing in developed countries, there is still little private sector biotechnology research effort in developing countries, particularly in Asia.

The commercialization and distribution of new agricultural biotechnology products, particularly transgenic crops, is also concentrated in Organisation for Economic Co-operation and Development (OECD) member countries, with a few exceptions. These products are for the most part crops of economic importance in industrial country agricultural production and in world trade in agricultural commodities, mainly soybean, maize, cotton, and canola.

During the past five years, the area under GMOs has increased rapidly from 1.7 million ha to 44.4 million ha, 75 percent of which are in the US. The remaining 25 percent are distributed in both developed and developing countries, including Argentina, Australia, Bulgaria, Canada, PRC, France, Germany, Mexico, Portugal, Romania, South Africa, Spain, Ukraine, and Uruguay. Relatively little biotechnology research is being undertaken on the problems of small farmers in rainfed and marginal lands. Neither is there much interest in Asia's basic food crops: rice, tropical maize, wheat, sorghum, millet, banana, cassava, groundnut, oilseed, potato, sweetpotato, and soybean. These are considered orphan crops because of the private sector's reluctance to work on them. That focus is unlikely to change because of the perception that investments in such orphan crops and from working on problems of small farmers yield limited returns.

To participate more fully in the biotechnology revolution, Asian governments will need to expand their capacities to undertake biotechnology research linked to the problems of small farmers and orphan crops. In certain situations, however, there may be opportunities to purchase, license, or import technology applicable in Asia.

Applications of Biotechnology

The applications developed from the new methods in biotechnology place them within the continuum of techniques used throughout human history in industry, agriculture, and food processing. Thus, while modern biotechnology provides powerful new tools, they are used to generate products that fill similar roles to those produced with more traditional methods.

There is now increasing use of modern molecular genetics for genetic mapping and MAS as aids to more precise and rapid development of new strains of improved crops, livestock, fish, and trees. Other biotechnology applications such as tissue culture and micropropagation are being used for the rapid multiplication of horticultural crops and trees. New diagnostics and vaccines are being widely adopted for the diagnosis, prevention, and control of fish and livestock diseases.

The science of genomics (the molecular characterization of all the genes in a species) has dramatically increased knowledge of plant genes and their functions. The new technologies enable greatly increased efficiency of selection for useful genes, based on knowledge of the biology of the organism and the role of specific genes in regulating particular traits. This will enable more precise selection of improved strains. These techniques may be used for more efficient selection in conventional breeding programs. They may also be used for the identification of genes suitable for use in the development of transgenic crops.

Thus far, scientists have completed genomic study on rice through the cooperative efforts of several international and private sector institutions led by Japan. Modem biotechnology permits increased precision in the use of new techniques and a shorter time to produce results. For example, plant breeders and molecular biologists can collaborate to transfer to a highly developed crop variety one or two specific genes to impart a new character such as a specific kind of pest resistance.

Table 2. Summary of Applications of Modern Biotechnology to Agriculture

Subsector	Applications
Crop Production Diagnostics.	To diagnose plant pests and pathogens, contaminants, and quality traits. Micropropagation techniques or tissue culture. To multiply disease-free planting materials on a large-scale.
Development of transgenic crops.	To develop commercially new genetically modified crop varieties. Modern plant breeding. To develop superior plant varieties rapidly and more precisely.
Marker-assisted selection	To use genetic markers, maps, and genomic information in breeding for high yielding, disease- and pest-resistant varieties.
Biodiversity	Characterizing, conserving, and using biodiversity.
Forestry Gene-mapping.	To accelerate tree breeding.

Macropropagation.	Rapid vegetative propagation by means of cuttings from large plantation of pines and other trees.
Micropropagation by tissue culture.	Large-scale multiplication of genetically superior plantlets. DNA finger printing. To differentiate species, strains, and cultivars accurately. Wood security. The selection of genetically superior trees for breeding purpose.
Livestock Production	Livestock improvement. To speed up the reproduction process in animals, allowing more generatiors to be produced.
Transgenic livestock	Development of transgenic lines of virus-resistant poultry and other animals.
Livestock health.	Application of diagnostics for the control of major diseases of livestock.
Vaccine development.	Development of vaccines for the control of epidemic viral diseases of livestock.
Fisheries Transgenic fish.	Still being explored. Use of molecular markers in biodiversity. Research, genomic mapping, and trait selection in fish and other aquatic organisms.

New techniques of modern biotechnology accelerate plant and animal breeding. They offer possible solutions to previously intractable problems and difficult targets such as drought tolerance, and enable the development of new products (Table 2). These products may include more nutritious food, crop varieties with improved tolerance for pests and diseases, and animal vaccines.

It is important to provide appropriate regulatory mechanisms to ensure that products produced by modern biotechnology are as safe as the products of traditional biotechnology. That is especially so when the products are GMOs that might interact with the environment. The knowledge of genes is not matched by our knowledge of the gene-environment interaction or potential impacts of biotechnology on the environment. However, many of the Green Revolution technologies were also introduced without such understanding. At present, there is widespread distrust of biotechnology and the public needs to be engaged in dialogue before the technology is disseminated widely.

Reducing Poverty Through Agricultural Science

Science and technology underpinned the economic and social gains in Asia over the past 30 years. In agriculture, these gains came to be known as the Green Revolution. Between 1970 and 1995, cereal produc-tion in Asia doubled, calorie availability per person increased by 24 percent, and real food prices halved. Although the region's population grew by 1 billion people, overall food production more than kept pace with population growth.

These food production increases were achieved largely by the cultivation of high-yielding varieties (HYVs) of rice and wheat, accompanied by expansion of irrigated areas, increases in fertilizer and pesticide use, and greater availability of credit. The scientific basis for the Green Revolution stemmed from national and international research programs that led to the development and distribution of new HYVs, particularly of rice and wheat. The first generation of these new varieties were based on the introduction of new genes for dwarfing that made the HYVs shorter, more responsive to fertilizers, and less prone to falling over or lodging when fertilized and irrigated.

Subsequent varieties also carried genes that gave increased pest and disease resistance and improved taste and grain quality. The key elements in improving food security in Asia from 1970-95 were government policies reflecting a belief that investments in increasing agricultural productivity were a prerequisite to economic development. These national policies were supported by political leaders in Asia and by both the public and private sectors of the international community.

This mix of supportive public policies, scientific discoveries, and public and private investments in rural Asia, particularly in irrigation, credit, and inputs, led to substantial reductions in poverty and improved food security throughout Asia over the past 30 years. Increased agricultural productivity, rapid industrial growth, and expansion of the nonfarm rural economy have all contributed to almost a tripling of per capita gross domestic product across Asia since 1970.

The intensification of agriculture and the reliance on irrigation and chemical inputs has led to environmental degradation, increased salinity, and pesticide misuse. Deforestation, overgrazing, and overfishing also threaten the sustainable use of natural resources. Green Revolution technologies had little impact on the millions of smallholders living in rainfed and marginal areas, where poverty is concentrated.

Furthermore, the Green Revolution has already run its course in much of Asia. Wheat and rice yields in the major growing areas of Asia have been stagnant or declining for the past decade, while population continues to increase. The key lessons learned from the Green Revolution are: (i) it has benefited farmers in irrigated areas much more than farmers in rainfed areas thus worsening the income disparity between the two groups, (ii) it overlooked the rights of women to also benefit from the technological advances, and (iii) it promoted an excessive of use of pesticides that are harmful to the environment.

As countries became self sufficient in food, government investments declined in the agricultural sector and in science and technology across the region. This reflects a worldwide trend toward declining public investments in the rural sector and in agricultural research and development (R&D), nationally and internationally. In Asia, private sector investments in the rural sector and related R&D have concentrated on export commodities. The downward trends in public investments by governments and development agencies in smallholder agriculture over the past decade have not been matched by a concomitant rise in private investments. Similarly, there is little (and few incentives for) private R&D on the food crops, livestock, fisheries, and aquaculture systems important for food security and poverty reduction in rural Asia.

Future Perspectives

The population of Asia is projected to increase from 3.0 billion to 4.5 billion in the next 25 years. During the same period, the urban population will nearly double from 1.2 billion to 2.0 billion, as rural people move to the cities in search of employment. These increases will place massive pressure on developing member countries (DMCs) of ADB to increase food production. Food demand is influenced by population growth, urbanization, income, and associated changes in dietary preferences. Urbanization and income growth frequently lead to shifts from a diet based on root crops (cassava, yam, and sweetpotato), sorghum, millets, and maize to rice and wheat, which require less preparation time, and to more meat, milk, fruits, vegetables, and processed foods. This dietary transition has already happened in much of the region.

Meeting the food needs of Asia's growing and increasingly urbanized population requires increases in agricultural productivity and matching these increases to dietary changes and rising incomes. To meet this demand, cereal production will need to be increased by at least 40 percent from the present level of about 650 million tons annually, most of which will have to come from yield increases. In addition, meat demand will double during the period.

Production increases will have to be achieved by increasing yields in a sustainable way to conserve diminishing and degraded natural resources. Nearly all of these production increases will need to take place in DMCs themselves because on average 90 percent of the world's food is consumed in the country where it is produced. Food imports are not only expensive but discourage the creation of employment, which is badly needed in the rural areas. In this millennium, we face a food, feed, and fiber production challenge in highly complex farming systems for several reasons:

(i) Water will become the most important limiting factor in agricultural production because the quality and quantity of water will decline as a result of pollution, forest degradation, and increased agricultural, domestic, and industrial use.

(ii) Urbanization will mean the loss of agricultural land to residential and industrial development, and a decline in the number of farm workers.

(iii) Most farmers are poor with small landholdings.

(iv) Farming systems are commonly heterogeneous with mixes of food crops, livestock, and trees.

(v) About 70 percent of the cultivated land is rainfed with unreliable distribution and intensity of rainfall.

Thus, the increase in food production during the next 25 years will have to be achieved using less labor, water, and cultivated land. This can be done only if scientists can develop new crop varieties with high yield potential and high water use efficiency. New understanding of plant and animal genes may offer ways to increase crop yields to the levels required to adequately and sustainably feed the growing population in Asia. .

Thus, developments in modern biotechnology could make extremely important contributions to future agricultural growth, food security, and poverty reduction. Increasing smallholder agriculture productivity will not only increase food supplies, but will reduce poverty and malnutrition, increase food access, and, improve living standards of the poor.

Biotechnology has the potential to (i) increase crop and animal productivity; (ii) improve nutritional quality; (iii) broaden tolerance of crops for drought, salinity, and other abiotic stresses; and (iv) increase resistance of crops to pests and diseases. These potential benefits will have significant impact in increasing food production and reducing poverty in DMCs if they can be applied to problems of the poor farmers in the tropics.

As with any technology, biotechnology brings with it potential risks. To maximize the benefits and minimize the risks, the introduction and use of biotechnology in DMCs or elsewhere must be thoughtfully managed by the public and private sectors. The key risks that relate to the application of new developments in biotechnology for the public good are food and environmental safety, economic concentration, and intellectual property (IP) management.

2

Agricultural Biotechnology

As in many other areas of technology, the introduction of a new technology such as agricultural biotechnology may depend on the perceived balance between the benefits of the technology and the potential risks to the environment and human health. Biology has been used for millennia to provide food and other products for use by humans. It is used to make food palatable or safe, and many mechanisms have been .discovered. that enable foods to be .modified. using organisms (usually micro-organisms). Bread, beer, wine, cheese, tofu and yoghourt are amongst the modified food products we use all the time.

The wild, undomesticated relatives of all food crops are almost unrecognisable in comparison to that which we grow in fields today. Genetic modification of crop plants has been practised since the crops were first identified. The first human to choose to keep the best seeds for growing in the following season rather than eating them was practising biotechnology.

Modern biotechnology is seen as different to traditional selection, for it permits the transfer of characteristics that could not be achieved naturally. Scientists often argue that the new techniques are simply an extension of the continuum of selection and genetic modification that has been used over hundreds of years. These traditional techniques have changed markedly during the 20 century as our understanding of the biological processes has improved. Deliberate mutation and many other artificial techniques have allowed selection of characteristics between weakly compatible organisms.

There are many scientists who believe that the transfer of genes between non-compatible organisms is truly different from traditional techniques and constitutes something that is really new. Modern biotechnology that permits modifications that cannot happen naturally has elicited excitement, fear and concern for many reasons, and has been regulated from almost the initial experiments that allowed genetic material to be transferred among unrelated organisms.

Different uses of modern biotechnology to produce transgenic organisms elicit varying reactions in most countries. Genetic modification to provide medicines is not as controversial as the genetic modification of crops for human consumption. Often the genetic modification of animals (especially reproductive cloning) is considered as less acceptable than modification of plants. Modification of the germ line in humans, for example, is usually considered as immoral.

Many opinion polls indicate that the public discriminates markedly between uses of biotechnology. Medicine and horticulture/floriculture are often found to be acceptable whereas the genetic modification of crops for food use, or the modification of animals and humans are less acceptable. .While most Americans say they would be in favour of at least some genetically modified food products, and nearly two-thirds believe that genetically modified foods will benefit many people, more than half (56 percent) say that the issue of genetic modification causes them great concern

Policy on the Safe use of Modern Biotechnology

Policy on the safe use of biotechnology set precedents. It is often the case that safety legislation is brought in because an accident has occurred and systems need to be put into place in order to ensure that it does not reoccur. The possible risks of modern biotechnology were recognised at the very beginning of its use, and steps were taken to ensure that it was used safely.

With the discovery (during the 1960s) of enzymes that could cut the DNA sequences at specific points in the sequence mechanisms became available for transferring information between organisms in a precise and controlled manner. If two molecules are cut using the same enzyme, they will be cut at identical sequences, and can therefore be wrongly re-assembled. It is thus possible to cut a piece of DNA and insert a desired sequence precisely at that point.

This technique was only initially effective in micro-organisms, as the insertion of specific molecules into the whole genome of plants is much more complicated. It only became possible when a bacterium was discovered capable of inserting a sequence of DNA into a random position in the plant genome. The ability to regenerate a complete plant from a single cell (impossible in other higher organisms) made it possible to easily modify many plant species. It was later discovered that if ballistics was used to literally fire DNA into cells a small proportion would have sequences stably integrated into their genome. Selection techniques involving (usually) choosing to have one of the inserted sequences express antibiotic tolerance mean that the tiny proportion of cells modified using the technique can be identified relative to the rest.

The potential uses of genetic modification were obvious from the moment that the techniques that enabled the transfer of genes from one organism to another unrelated organism were first identified. A United Kingdom government initiated committee reported in 1975 that genetic manipulation techniques would provide .substantial though unpredictable benefits. . .application of the techniques might enable agricultural scientists to extend the climatic range of crops and to equip plants to secure their nitrogen supply from the air.

A meeting of scientists using the very new recombinant DNA technology in Asilomar in California in February 1975 produced a set of guidelines for the use of biotechnology: The formal goals of the meeting included the need to identify the .possible risks involved for the investigator and or others. and .the measures that can be employed to test for and minimise the biohazards so that the work can go on. The benefits of the new technology far outweighed the risks if suitable precautions were put in place in the view of the Ashby committee in 1975.

The use of modern biotechnology, however, extends far beyond arable agriculture, and it is heavily used in the industrial manufacture of high value chemicals and pharmaceuticals. In addition animals are being modified; higher animals acting as factories or to use them more efficiently for food; insects for research purposes and to attempt to eliminate disease, etc. For safety reasons the contained or confined use of genetically modified organisms is likely to be favoured over crop plant modifications particularly where they are used as factories for the manufacture of toxic chemicals and pharmaceuticals.

Transgenic organisms may be used for human and animal medicine that are then purified for chemicals rather than sold as organisms. Transgenic viruses may be used as vaccines through to vectors for inserting needed gene sequences into other organisms. Bacteria and fungi are likely to be used industrially for making chemicals. It is likely that these modified microorganisms will be traded to be used industrially, or used in factories to produce wanted chemicals.

Modified micro-organisms are already being used for all sorts of purposes, and are likely to be used directly as bio-control agents in the future. Small animals. (insects for example) may be manufactured either because they are the only organisms identified for a particular use, or because they are a perfect vector for a particular reason e.g. mosquitoes incapable of carrying the malaria parasite. In most instances these will be released into the environment and will impact on the ecology into which they are released.

Large animals are being used as factories for the production of high value chemicals. Animals that express pharmacologically active compounds in their milk, for example, are contained and effective factories . but care has to be taken to ensure that the animals are treated humanely and are not seriously affected because of the nature of the chemical they are being used to fabricate.

Although the public in many countries has been fearful of the introduction of the .products of this technology, Parliaments have not been as reticent, and have recognised both benefits that may arise from its use and the risks that it theoretically poses. In 1993 the Parliamentary Assembly of the Council of Europe passed recommendation 1213 on developments in biotechnology, for which there were many wonderful prospects, but also for which there were many concerns. The Council of Europe includes many countries in Central and Eastern Europe as well as those of the affluent European Union. The resolution noted that the gene pool has been widened far beyond the limits of sexual compatibility to encompass the possibility of transferring genes from almost any organism to others.

Amongst the many uses of biotechnology it identified were the raising of agricultural outputs (or reducing inputs), the replacement of chemical herbicides and insecticides or more efficient targeting, the use of plants in industry, changes in responses of crop plants to stress and even the cloning of meat animals .for particular markets or to form embryo banks to maintain genetic diversity.. The resolution noted that there might be significant drawbacks resulting from the application of the new biotechnology. The possibility of new diseases was raised, as were the potential environmental effects of transgenic organisms. Many of the benefits have been effected, although many do not realise that vaccines, pharmaceuticals and food additives (such as chymosin and ascorbic acid) are often the products of modern biotechnology.

The Cartagena Protocol was agreed in 2000 after years of negotiation and argument, with many misgivings, but in an atmosphere which had changed from that which pertained at the time the negotiations started. Article 19(3) of the Convention on Biological Diversity had required parties to consider the possibility of a Protocol to the convention that addressed the use (and primarily transboundary movement) of living modified organisms that might have an adverse impact on biological diversity.

Eight years later Europeans were no longer accepting modern biotechnology; products had disappeared from the shops, and there was a gloom and distrust in many countries not observed elsewhere. Few if any products derived using modern biotechnology are now available in Europe. In North America, farmers adopted transgenic organisms with little opposition, and products derived from them have been in the shops for over 5 years.

The Developing Countries had wanted far more in the Protocol than they were able to get, with many more safeguards. The producer countries fought hard to ensure that, insofar as it was possible, few if any controls were applied particularly to commodity goods. The size of the commodity market alone, they argued, made it difficult to contemplate a regime which required what amounted to .visas. at country entry points.

The Protocol requires 50 ratifications to come into force. To date it has received 45 ratifications, one of which is by the European Union, which does not count as part of the 50 required (not all the European Union countries have ratified . when they do, the number will exceed the threshold). More than 100 countries have signed. Southern African countries that have signed but not yet ratified are Malawi, Namibia, Uganda and Zimbabwe.

Amongst those which have ratified or acceded to the protocol are Mozambique, Botswana, Mauritius, Kenya, Uganda and Lesotho. Most of these countries do not yet have the legal systems in place to implement the requirements of the Protocol. The major user of transgenic organisms in southern Africa, South Africa, was involved in the discussions during its adoption but has not (yet) signed or ratified.

The need for specific legislation in regard to the use of genetically modified organisms was never a presumption even though it was recognised that regulation was needed from the earliest days of the use of this technology. The UK had regulated the genetic .manipulation. of micro-organisms starting in 1978, and by 1983 had a full set of legally binding regulations in place. The United States, on the other hand, had specified guidelines (the NIH Guidelines) which identified the manner in which such organisms should be used by those funded by the National Institutes of Health.

In 1986 the US Government published its Coordinated Framework for the Regulation of Biotechnology which described the comprehensive federal regulatory policy for ensuring the safety of biotechnology research and products. Existing statutes provide a basic network of agency jurisdiction over both research and products; this network forms the basis of this coordinated framework and helps assure reasonable safeguards for the public. This framework is expected to evolve in accord with the experiences of the industry and the agencies..

The laws that already existed in the United States regulated specific product uses, such as foods or pesticides. It was considered that genetically modified organisms posed no new risks that could not be covered using the existing system. This approach provides the opportunity for similar products to be treated similarly by particular regulatory agencies. The underlying policy question was whether the regulatory framework that pertained to products developed by traditional genetic manipulation techniques was adequate for products obtained with the new techniques.

Upon examination of the existing laws available for the regulation of products developed by traditional genetic manipulation techniques, the working group concluded that, for the most part, these laws as currently implemented would address regulatory needs adequately. For certain microbial products, however, additional regulatory requirements, available under existing statutory authority, needed to be established.

The US Administration decided to identify the various tasks needed to regulate biotechnology and clearly indicate the Agency and even the law which would be used to ensure that these technologies were used safely. Other countries did not (at the time) have the range of environmental, food, drug and safety legislation in place that permitted effective use of existing legislation. In the US it was decided that jurisdiction over the many different biotechnology products would be determined by their use rather than the manner of their products, just as was the case for traditional products.

Regulatory Systems

Guidelines or regulations were quickly introduced in some countries, particularly to protect those who might come into contact with the modified organisms. In the UK the first regulations were introduced in 1978; the NIH guidelines were implemented in the United States soon after Asilomar and applied to work funded through grants received from the NIH. Initially the .regulations. applied primarily to work in laboratories because that was the only place in which the work could progress. They were aimed at the protection of those individuals who had access to the laboratories or facilities and attempted to assure that the work was contained and that workers were protected from the hazards posed by the modified organisms.

It was only in the late 1980s that the introduction of modified organisms into the environment became really feasible. At first it was expected that these releases would mainly be of micro-organisms, but as methods capable of modifying plants became available and efficient it was clear that most environmental releases would be of plants. Very few modified microorganisms have been released. Many countries have decided to implement different systems of regulation for organisms intended for use in containment (or confined) and those that are released into the environment. Contained use included organisms used in industrial plant and processes for manufacturing where the organisms themselves are not intended to be marketed or exposed to the .open. environment.

Most countries in the Southern African Region are considering the frameworks necessary for a regulatory system to ensure the safe use of modern biotechnology or have already enacted legislation. South Africa initially regulated transgenic organisms through a voluntary system but since 1997 have had legislation in place to ensure that the use of modified organisms in South Africa is done safely. It is the only country in the region that has, so far, permitted the commercial use of any transgenic plants.

Zimbabwe was the second country after South Africa to come up with bio-safety regulations; was the first to come up with an institutional framework and is one of the few countries to have post-graduate training in biotechnology.. Namibia was part of a UNEP/GEF financed Pilot project which permitted 18 countries to start the process of regulating Biotechnology and is now one of 12 countries financed by the GEF to implement the biosafety frameworks that have been devised for the country.

Zambia, Uganda and Kenya were also among the countries that participated in the Pilot project and Uganda and Kenya are among the twelve now implementing their frameworks with significant funding from the GEF. Botswana, Lesotho, Mozambique, Rwanda, Zimbabwe and others in the region are currently being funded through a follow up project to the Pilot which assists countries to design frameworks to assure safety of biotechnology Countries have chosen to use a variety of triggers for regulation of biotechnology.

In Europe it is the fact of using modern biotechnology as defined in the Directives that triggers the regulatory process. In the United States, because current law is used, the trigger tends to be the use of organisms that are pests . plant pests for example - in the manufacture of the organism if the Department of Agriculture is to be involved. Canada has chosen to use a concept of novelty to trigger the regulatory process. Many analyses have suggested that once the process is started, the risk assessment and management processes are very similar.

Environmental Policy in relation to GMOs

All the countries that are participating in GEF funded projects have signed the Cartagena Protocol on Biosafety that specifically requires regulation in relation to the trans-boundary transfer of living modified organisms that may have adverse effects on the conservation and sustainable use of biological diversity, taking also into account risks to human health. Those participating in the implementation. projects have also ratified or acceded to the Protocol or have agreed to do so.

They are also all party to the Convention on Biological Diversity which requires, through Article 8(g), that they institute national frameworks in order to establish or maintain means to regulate, manage or control the risks associated with the use and release of living modified organisms resulting from biotechnology which are likely to have adverse environmental impacts that could affect the conservation and sustainable use of biological diversity, taking also into account the risks to human health.

The provisions of the Cartagena Protocol extend only to those organisms resulting from modern biotechnology that might cause potential adverse effects to the conservation and sustainable use of biodiversity. Human health has .then. to be taken into account. However, when designing a regulatory system for biosafety, it is legitimate to assure safety of the environment and human health in general, with the needs for the Protocol forming a sub-set within the regulatory system. It seems likely that any attempt to link the protection of human health to legislation that primarily addresses biodiversity would not be acceptable to most legislatures.

Countries have understood that biosafety in this instance means primarily protection of the environment, and that the release of living modified organisms needs be regulated in order to protect the environment. Safety concerns are not, however, limited to the impact on the environment and regulatory systems that attempt to assure human and animal health are often different from that set in place for environmental protection. The European Novel Food Regulation provides extensive risk assessment and management for the use of genetically modified organisms or products derived from them in foods. Amongst other foodstuffs, it applies to:

(a) foods and food ingredients containing or consisting of genetically modified organisms within the meaning of Directive 90/220/EEC;

(b) foods and food ingredients produced from, but not containing, genetically modified organisms;.

Precaution

Scientific data is able to be collected at many sites around the world which can provide an insight into the manner in which the product may interact with its environment when released into a particular environment. Where data is not available or where a country believes that its environment is different from that in which the organism was tested it may require field testing before an organisms is released or placed on the market. Where data is knowable further experimentation will provide data that may address concerns and provide answers as to the likely behaviour of the organism in a particular environment. However, because of the inherent variability of biological systems information as to the behaviour may fall into the .not-knowable. category where no amount of information collected can provide more than increased precision in determining the variability.

The precautionary approach is invoked in order to address the absence of data. It is usually taken to refer to Principle 15 of Agenda 21 agreed in Rio de Janeiro in 1992. In order to protect the environment, the precautionary approach shall be widely applied by States according to their capabilities. Where there are threats of serious or irreversible damage, lack of full scientific certainty shall not be used as a reason for postponing cost-effective measures to prevent environmental degradation..

There have been many cases of serious environmental degradation that have made governments change their perception of environmental protection. These have also affected the public.s perception of the environment. Outbreaks of disease in animals and humans due to lack of care (as it is perceived) or to environmental pollution has had a significant effect on an appreciation of both known and potential risks to the environment and to human health and a public acceptance that these potential problems need to be addressed.

The use of precaution cannot be limited to approving an action or process, or prohibiting it, but implies managing various levels of risk and uncertainty, and taking the appropriate measures at each level. A risk may vary significantly depending on the level of activity or the likelihood that an organism may persist and establish itself in the environment. The inter-relationship with other actins or processes or other organisms with which genetic material may be exchanged may also require caution in analysing the potential risk.

This precautionary principle (or approach) has attracted many and various interpretations; for many it means that if the science is unknown and there is a risk of environmental damage, do not proceed. Caution dictates that it implies that where there is doubt over the safety of an action, that action should not be taken until evidence is available that the steps to be taken will not have disastrous consequences for the environment.

The concern in relation to transgenic organisms is due to the possibility that once an organism is in the environment it is virtually impossible to recall and because of its property of replication, it does not decay over time; indeed numbers may increase disastrously. Others interpret this as an injunction to proceed with caution, considering each release into the environment on a case-by-case basis and probably also proceeding step-by-step, where small field trials precede larger ones and the results are analysed before proceeding to commercial unfettered release .

Recourse to the approach .presupposes that potentially dangerous effects deriving from a phenomenon, product or process have been identified, and that scientific evaluation does not allow the risk to be determined with sufficient certainty. Decision making about risks in the context of a precautionary approach is further complicated by the inherent dynamics of science. Even though scientific information may be inconclusive, decisions will still have to be made to meet society.s expectations that risks be addressed and living standards maintained. Scientists may be concerned that the .principle. is used to stifle research, innovation and competition.

Where action is deemed necessary, measures based on the precautionary principle should be, inter alia:

— proportional to the chosen level of protection,

— non-discriminatory in their application,

— consistent with similar measures already taken, based on an examination of the potential benefits and costs of action or lack of action (including, where appropriate and feasible, an economic cost/benefit analysis),

— subject to review, in the light of new scientific data, and capable of assigning responsibility for producing the scientific evidence necessary for a more comprehensive risk assessment..

The use of precaution requires a number of major considerations to be taken into account. The Canadian Principles provide a starting point for defining policy in relation to precaution.

1. The decision-making process for managing risks always requires sound and rigorous judgment. where [J]udgment means determining what is a sufficiently sound or credible scientific basis, what follow-up activities may be warranted, and who should produce a credible scientific basis.

2. To reduce significant scientific uncertainty and improve decision making, the precautionary approach usually includes follow-up activities such as research and scientific monitoring..

However, it has to be noted that in many instances the collection of data may increase the precision of determination of variation, rather than provide data which permits the reduction of uncertainty. Monitoring can only provide assurance that expected events occur, and events predicted not to occur are not observed.

Unexpected, unpredictable, indirect and delayed effects on the environment are by their nature difficult if not impossible to monitor. The arguments that have ranged around precaution are serious, for they have directly impacted on the policy decisions of many countries. In Europe precaution in relation to transgenic organisms is taken to require case-by-case and step-by step approaches torisk. This way of interpreting precaution is built into the Protocol, which also requires a case-by-case process in assessing risk.

Public Opinion

The controversy over the use of modern biotechnology has centered primarily on commercial release into the environment, rather than use in laboratories for research, contained use in industry, use in the production of pharmaceuticals and veterinary products or even field trials. Protesters have, however, chosen to attack and destroy fields in which organisms are being tested. It is the industrial use of genetically modified organisms that may be the major use of modern biotechnology now and in the future.

The Eurobarometer surveys show that there is considerable discrimination amongst the public (at least in Europe) in relation to the various uses that modern biotechnology can be put. .Europeans continue to distinguish between different types of applications, particularly medical in contrast to agri-food applications.. For GM crops and GM foods support declined and opposition increased over the period 1996-1999, from 1999-2002 there is almost no change in levels of support or opposition. European attitudes to six applications of biotechnology indicate the discrimination that has been observed.

In 1994 approximately 7,000 acres were planted under 593 USDA field-test authorisations, compared to 57,000 acres under 1,117 authorisations in 2001. The first biotechnology-derived crops were commercialised in 1996 and, in 2001, approximately 88 million acres were planted in the United States and 130 million acres were planted world-wide. Canada, Argentina and Mexico are the only other countries in which there has been significant use of modern agricultural biotechnology, although many other countries including South Africa and Australia are starting to increase their use of living modified organisms in agriculture.

China has approved a small number of transgenic varieties of cotton and expects to proceed to the commercial production of modified rice in the next two years. The latest Eurobarometer survey of European attitudes to biotechnology has indicated a recovery of faith in technology, but this may simply be that the de facto moratorium on the commercialisation of plants manufactured using GM techniques has taken the subject out of the public consciousness:

In the United States the American public's position on the acceptability of genetic modification of food is decidedly . . . undecided.. 58% of Americans either strongly approve or somewhat approve of creating hybrid plants using genetic modification, while 37% disapprove. Many developing countries are fearful of the impact of agricultural biotechnology. Zimbabwe and Zambia, for example, have been wary of permitting food-aid which contains transgenic maize into the country, even though many of its people are starving.

The UNEP International Guidelines and the Cartagena Protocol require the public to be informed and educated about biosafety, but the virulent reaction against the technology in Europe impacts directly on any public image more easily than a reasoned argument for the safe use of the technology. During the first nine months of 1999 in Britain there were a continual series of press reports implying that eating GM food would lead to all sorts of serious diseases..

The attention paid by the media to foods produced using modern biotechnology has been sustained over a long period and is almost totally hostile. The coverage has stressed the technology, rather than the products. The rejection of genetically modified foods by many European Supermarkets and food producers has impacted on production and growing of genetically modified crops that have to be exported to one of the largest food markets in the world. To grow rice modified so that it produces vitamin A is a wonderful prospect for nutrition in the many countries that depend on rice as a primary food.

GM crops are irrelevant to ending hunger; the new technology puts too much power over food into too few hands; and too little is done to help small farmers grow food in sustainable and organic ways. It is tempting to see biotechnology in agriculture as a clean neutral science, simply transferring progress from the laboratory to the field, improving the lot of everyone. All technologies are embedded in specific economic and social systems and have different costs and benefits. This response in Western Europe to the new technology cannot easily be dismissed through assertions by scientists that there is negligible risk, or that permits to market transgenic foods and crops (in particular) should be based solely on risk assessments that are science based.

If all the scientific information had been available and a consensus amongst scientists could be achieved that the impact on the environment is minimal, it would be possible to argue for a totally science based risk assessment process. There are many different ways in which policy on involving the public has evolved. The Cartagena Protocol requires that countries engage their public in decision making both at the policy level and when considering individual applications for use of modern biotechnology. For countries with a history of involving their public in the decision making process this is not easy; for those not used to direct public involvement it may be much more difficult.

The Parties shall:

(a) Promote and facilitate public awareness, education and participation concerning the safe transfer, handling and use of living modified organisms in relation to the conservation and sustainable use of biological diversity, taking also into account risks to human health. In doing so, the Parties shall cooperate, as appropriate, with other States and international bodies;

(b) Endeavour to ensure that public awareness and education encompass access to information on living modified organisms identified in accordance with this Protocol that may be imported.

The Parties shall, in accordance with their respective laws and regulations, consult the public in the decision-making process regarding living modified organisms and shall make the results of such decisions available to the public, while respecting confidential information in accordance with Article 21.

Science-based Decisions

There have been many arguments that decisions on the use of these living modified organisms must be based on science; Policy may be defined when designing the system that is applied to individual applications, but that the applications should only be considered in the light of this policy. Decisions are usually made by Government based on advice received from a number of sources. The risk assessment procedure, at the very least, should be science based. This is made very clear in the Cartagena Protocol. Article 15 states that .Riskassessments undertaken pursuant to this Protocol shall be carried out in a scientifically sound manner..

The Parties, in reaching a decision on import under this Protocol or under its domestic measures implementing the Protocol, may take into account, consistent with their international obligations, socio-economic considerations arising from the impact of living modified organisms on the conservation and sustainable use of biological diversity, especially with regard to the value of biological diversity to indigenous and local communities.

Unlike the European Union, the US and Canada, the vast majority of developing countries may not have expertise directly employed by government in the vast array of disciplines needed to perform a complete risk assessment for transgenic organisms. The data needed to assess likely environmental degradation or impact may not be available in many countries. A different approach may be needed, where an applicant requesting a permit for the use of a transgenic organism must perform a detailed risk assessment possibly even performing field tests in an appropriate environment . and submit this for audit to a government, rather than the government performing the risk assessment.

Most scientists may feel more confident at auditing a detailed assessment than attempting the assessment themselves. Applicants could also be expected to design their own risk management, consultation and monitoring procedures, with input from government appointed assessors where appropriate. There is an obvious danger in this approach, however, for lack of trust in those applying to release organisms to provide all the necessary information may mitigate against the acceptance of the risk assessment. Can applicants be trusted to provide all the necessary information? Where a decision is made to use audit rather than direct risk assessment by government, it is important that the scientists involved in the audit are able to ask for further information and are able to identify gaps in the approach taken by the applicant.

Risk assessment of genetically modified organism is largely based on the concept of familiarity, or of .substantial equivalence.. It is actually difficult to identify other ways of approaching the problem of identifying risk. It assumes that the characteristics of the modified organism except for the specific characteristics introduced are those of the host organism. It is not easy to see any other approach, but the Royal Society of Canada and the Royal Society in the UK have both indicated dissatisfaction with .substantial equivalence. which is arguably substantially equivalent to familiarity.

Crop varieties developed through conventional plant breeding techniques not involving modern biotechnologies are not generally tested for their safety. Rather they have to meet plant variety registration requirements that identify whether they are distinct from those currently on the market, uniform and stable. These traditional methods use (primarily) crossing selection and back-crossing processes to select a desired characteristic and remove inadvertently introduced extra characteristics that initially accompany the introduced trait. These mechanisms introduce new and numerous gene combinations. If there are toxins known to occur in these crops (e.g. glucosinolates in canola, glycol-alkaloid accumulation in potatoes), the new variety is normally tested to ensure that the level of toxin (or allergen) is no greater than the range that is normally observed for that substance.

Interactions of traits introduced by traditional methods with other characteristics of the plant are normally ignored until they prove to make the variety unusable. The implicit assumption behind this methodology is that, even where a breeding-derived novel trait is involved, new combinations of existing genes operating within highly selected germplasm are not expected to generate harmful outcomes. For transgenic crops, the concept of .substantial equivalence. was introduced. If a new food or food component is found to be substantially equivalent to an existing food or food component, it can be treated in the same manner with respect to safety.

In 1995, the World Health Organisation published a report in which the concept of substantial equivalence as a decision threshold was promoted as the basis for safety assessment decisions concerning GMOs. Substantial equivalence can be considered in three ways:

— the GM foodstuff might be regarded as substantially equivalent to its conventional counterpart both toxicologically and nutritionally. When a product has been shown to be substantially equivalent, no further safety assessment is required.

— it might be substantially equivalent apart from certain defined differences. Sometimes the GM food product includes the components deliberately introduced by genetic modification. In this case the GM food product might be regarded as substantially equivalent to its conventional counterpart except for a small number of clearly defined differences. Assessment is then limited to examining the implications of the difference(s), perhaps by testing the novel components of the GM plant in isolation..

— the GM product might be regarded as not substantially equivalent to its conventional counterpart, or there might not be a suitable reference available for comparison. The product will then need a highly detailed safety assessment. taking all the properties of the modified foodstuff and determining by direct measurement where necessary the impact on human health and the environment.

Many countries are deciding that the use of the term substantial equivalence. is misleading. It suggests that if substantial equivalence is demonstrated, no further assessment need be done. There is a mistaken perception that the determination of substantial equivalence was the end point of a safety assessment rather than the starting point..

Intellectual Property Rights and Ethics

There are many arguments for and against the use of intellectual property rights in relation to modern biotechnology. The patent system, as a system for the protection of intellectual property, is an integral part of the market economy and therefore can be a driving force for innovation in many technological questions. The same resolution notes that living organisms are able to reproduce themselves even if they are patented, and in view of this special quality of living organisms the scope of a patent is difficult to define, which makes it nearly impossible to find a balance between private and public interests.

Over the last few decades the global trading importance of biotechnology has been recognised and as a result there have been concerted and concentrated efforts to protect the results of research and development involving genetic material. The result of this has been the extension of intellectual property protection to most forms of biological material. The trade importance of biological information has been underlined the adoption of the Agreement on the Trade Related Aspects of Intellectual Property Rights (TRIPS) within the World Trade Organisation. This requires states party to the agreement to provide protection for all types of inventions irrespective of the field of technology.

The aim of the Agreement is to ensure that all member states provide effective and appropriate intellectual property protection and protect the intellectual property rights by the appropriate enforcement mechanisms. The Agreement sets down the minimum standards of protection. The TRIPS agreement permits countries to exclude from patentability those inventions whose commercial exploitation may be contrary to ordre public or morality. Countries may exclude from patentability .diagnostic, therapeutic and surgical methods for the treatment of humans or animals. and more importantly.

Article 27(2) allows Members to exclude from patentability innovations in order to protect animal, plant life or health or to avoid serious damage to the environment and Article 27(3) provides for exclusion from patentability of .plants and animals other than micro-organisms, and essentially biological processes for the production of plants or animals other than non-biological and microbiological processes. However, Members shall provide for the protection of plant varieties either by patents or by an effective sui generis system or by any combination thereof..

What constitutes sui generis protection for new plant varieties is not defined hence countries are free to adopt a system that ensures intellectual property protection for plants. One option is for countries to implement the International Convention for the Protection of New Varieties of Plants (UPOV) but this form of protection has been criticised for focusing too much on the rights of plant breeders, and too little on the rights of those using the seeds . farmers.. Whilst obliging member states to provide protection systems, those .inventing. new products do not need to obtain that protection. The rights only apply in the country in which the inventors have chosen to invoke protection.

In most African countries many of the biotechnology inventions will not have been protected through patent rights and can legally be used as if in the public domain. It is only when products developed using patent protected materials or methods are exported into countries where protection exists that the rights of the inventor must be respected. There is an underlying assumption that the introduction of an intellectual property system will result in a dramatic increase in the innovative capacity of the private sector, while allowing the public sector to become more self-financing. This may be true to an extent in countries with a substantial research capacity but is unlikely to be the case in developing countries where the research and development sector is not as strong.

A Northern intellectual property system may provide an incentive but there may be limited local capacity to exploit it. Even when technologies are developed, firms in developing countries can seldom bear the costs of acquisition and maintenance of rights and to an even greater extent, enforcement (especially in those countries where substantial earnings may be realisable). The costs of establishing an infrastructure to support a intellectual property rights regime may be substantial, and mechanisms for the enforcement of IP rights is costly both to government and to private stakeholders.

If there is a policy commitment to the implementation of a rights system, then perhaps the best way to proceed would be to look at the systems in Europe and the United States and adapt them to local and cultural needs. The required patent system would need to balance the costs and benefits to local needs and requirements. There is real concern about the use of intellectual property law in developing countries, particularly in relation to health care but also in relation to that which is emotively called biopiracy or bio-prospecting.

Pharmaceutical companies worldwide are interested in finding new and alternative therapies and have widened their search to include traditional medicines and practices largely based on medicinal plants endemic to developing countries. There are many traditionally used herbal medicines which may have real therapeutic properties. If a company takes the knowledge as the starting point for a serch for new pharmaceuticals, extracts the active product it is entitled to a patent on the extracted product even though it cannot replace the traditional product itself. Developing countries are thus faced with the acute dilemma of their valuable indigenous wealth being taken away and exploited commercially by the resource and technology rich trans-national pharmaceutical companies.

The International Center for Tropical Agriculture (CIAT) is legally challenging the patent, arguing that the patent claims are invalid, failing to meet novelty and non-obviousness requirements and disregarding available prior art. In the US under Section 35 U.S.C. §102(a) the invention cannot be .known or used in this country, or patented or described in a print publication (emphasis added) in this or a foreign country.. Therefore, mere use in Mexico without printed publication is insufficient to show a lack of novelty.

Membership of the WTO requires countries to have in place an effective intellectual property regime. However the simple implementation of the Agreement into national law is insufficient to protect a country.s genetic resources as Article 27(3b) is inadequate to meet their protection requirements. What is required is the enactment of legislation that incorporates the framework of current agreements and negotiations . TRIPS, the requirements of the Convention on Biological Diversity and the International Treaty for the Protection of Plant Genetic Resources.

Hence Developing countries should consider the manner in which they implement the various agreements in order to protect their people and their resources, including

1. Enacting appropriate biodiversity protection legislation including benefit sharing consistent with Article 8j of the CBD and access to genetic resources (Article 15)

2. The TRIPS agreement does not require countries to institute a patent regime for plant material but replace it with a sui generis system for protection of the plant intellectual regime. The replacement system could be designed to protect extant varieties that are in the public domain as well as new plant varieties and provide for the needs of the country taking into account, for example, the communitarian approach to property that are often part of the culture of developing countries as well as the needs for innovation.
3. Developing countries may need to document and catalogue their biological assets not only to ensure protection but also to assure future collaboration and exploitation. States have sovereign rights over their biodiversity and are responsible for conserving their biological diversity and for using their biological resources in a sustainable manner. Article 3 of the Convention of Biological Diversity states that .States have, in accordance with the Charter of the United Nations and the principles of international law, the sovereign right to exploit their own resources pursuant to their own environmental policies, and the responsibility to ensure that activities within their jurisdiction or control do not cause damage to the environment of other States or of areas beyond the limits of national jurisdiction.

There are many who are concerned at the way in which IP protection has been used in many countries. The balance in many IP systems seems to be shifting too far in favour of technology producers. Negotiations over IPRs have been powerfully influenced by industry lobby groups and are being driven by concerns of trade liberalisation and international investment between developed countries. The legitimate technological and developmental objectives of developing countries generally technology users are not being given due consideration. This shift in the ownership and control of information, and the resulting boon to private investors, has been called an .information land grab.

Economic and Social Benefits and Risks

Of necessity biotechnology has to be applied for the benefit of human beings, society and the environment. These are not necessarily the same, for the benefit to living human beings may be at the short or long term cost to the environment. There is a presumption that .acceptability of the risk must include an improved quality of life, perhaps as we develop better (or more) food, better health and an environment which is improved in a sustainable manner.

Human usage of the environment in the 10000 years of our exploitation of nature has been relatively benign. In the last 100 years, however, we have made rapid and possibly irretrievable changes to the environment, including the excessive use of fossil fuels relative to their replacement, excessive use of water, production of greenhouse gases etc and even a huge increase in the human population. Humans are no longer in harmony with their environment, and we have to be aware of an impact on the environment. Where a primary goal was the pursuit of happiness (and the greatest good) we now have to pursue sustainability. These concerns are human centred. Many of those that live in Southern Africa are suffering from severe malnutrition, and drought is causing havoc with and to the environment.

If the application of modern biological techniques could result in food products that can better survive drought and heat, and also is able to provide more food in the right place and the right time, then there are clear benefits that result from its use. It is axiomatic that food is essential for our survival. .Both formal ethical systems and ethical practices in every society presume the necessity of providing those who are able-bodied with the means to obtain food and enabling those who are unable to feed themselves to receive food directly.

We consider it intolerable that more than 800 million people throughout the world, and particularly in developing countries, do not have enough food to meet their basic nutritional needs. This situation is unacceptable. Food supplies have increased substantially, but constraints on access to food and continuing inadequacy of household and national incomes to purchase food, instability of supply and demand, as well as natural and man-made disasters, prevent basic food needs from being fulfilled. The problems of hunger and food insecurity have global dimensions and are likely to persist, and even increase dramatically in some regions, unless urgent, determined and concerted action is taken, given the anticipated increase in the world's population and the stress on natural resources..

It is clear that we need to promote access to the genetic resources for food and agriculture for farmers, farming communities and the consumer. Health is improved where hunger and quality of food is eliminated. Healthy people are empowered in that they are able to participate in society and are more able to live meaningful lives. The FAO constitution identifies the need to raise levels of nutrition, secure improvements in the efficiency of production and distribution of all food and agricultural products, and better the conditions of those who live in rural areas.

For most consumers in developed countries the choice of whether to eat genetically modified foods or not is not an ethical issue. To eat genetically modified food would not be wicked, even if the individual was concerned as to its safety. If however, that food was proscribed by the society as (for example) not being halal or kosher, then an inability to identify the food as proscribed would be unethical. Where there are people who are starving and where a technology can help to provide for more and nutritionally better food, and it is not made available is an ethical issue.

The industrialisation of agriculture is an issue in many African countries, for it takes away the traditional structures of society and substitutes a more individualist system that may cause harm. This industrialisation may arguably help in providing more and better food at the cost of disrupting traditional belief systems and modifying the way of life of many in rural areas which may result in less food available where and when necessary.

The agreement setting up the World Trade Organisation tries to balance the many conflicting issues that this principle requires - relations in the field of trade and economic endeavour should be conducted with a view to raising standards of living, ensuring full employment and a large and steadily growing volume of real income and effective demand, and expanding the production of and trade in goods and services, while allowing for the optimal use of the world's resources in accordance with the objective of sustainable development, seeking both to protect and preserve the environment and to enhance the means for doing so in a manner consistent with their respective needs and concerns at different levels of economic development. The WTO and its disputes resolution system has placed the freedom to trade above environmental concerns, but there is recognition of the importance of environmental concerns.

Sustainable Development

In 1987 the Brundtland Report, also known as Our

Common Future, considered the need for ensuring that economic development was achieved whilst natural resources were not depleted. It is, the report asserted necessary to provide for the future without harming the environment. It is in the hands of humanity to make development sustainable, that is to say, seek to meet the needs and aspirations of the present without compromising the ability of future generations to meet their own.

The concept of sustainable development implies limits -not absolute limits, but limitations that the present state of technology or social organisation and the capacity of the biosphere to absorb the effects of human activities impose on the resources of the environment-, but both technology and social organisation can be organised and improved so that they will open the way to a new era of economic growth. The Commission believes that poverty is no longer inevitable. Poverty is not only a malaise in itself. Sustainable development demands that the basic needs of all are satisfied and that the opportunity of fulfilling their expectations of a better life is extended to all. A world where poverty is endemic will always be susceptible to suffering an ecological or any other kind of catastrophe.

3

Plant Biotechnology to Agriculture

Plant biotechnology complements plant breeding efforts by increasing the diversity of genes and germplasm available for incorporation into crops and by significantly shortening the time required for the production of new cultivars, varieties and hybrids. From an economic perspective, plant biotechnology offers significant potential for the seed, agrochemical, food processing, specialty chemical and pharmaceutical industries to develop new products and manufacturing processes.

Perhaps the most compelling attribute of the application of plant biotechnology to agriculture is its relevance both to helping ensure the availability of environmentally sustainable supplies of safe, nutritious and affordable food for developed countries; and to providing a readily accessible, economically viable technology for addressing primary food production needs in the developing world.

The need for new agricultural technologies, in general, is driven by two distinct, and at times contradictory societal requirements-ensuring a safe, nutritious, and affordable food supply for the planet, and at the same time, minimising the negative environmental impacts of food production itself. It is estimated that world population will double in the next 40 years, to exceed 10 billion. The combination of population increase, the decline in the availability of arable land, and the need for improvements in the quality of dietary intake in many developing countries means that agricultural production will have to be doubled, or even tripled, on a per acre basis to meet this need.

At the same time, societal concerns over the environmental impact of certain agricultural practices will increasingly restrict the types of tools that can be used in crop production. How will agricultural systems evolve by the year 2030 to meet these needs? How do we increase the productive efficiency of existing cultivated land without irreversibly damaging the planet? The answer is deceptively straightforward: investment in, and development of, new agricultural technologies is absolutely critical for a sustainable agriculture for the future.

Current agricultural technologies such as plant breeding and agrochemical research and development (R&D) will continue to play a major role in assuring a plentiful and safe food supply; environmentally sensitive and economic farm management practices will also play an important role. Advances in all these areas will be required to meet world food production needs. Plant biotechnology is uniquely important in this regard because it is:

— a new tool which can significantly impact crop productivity;

— compatible with sustainable, environmentally sound agricultural practices;

— a non-capital intensive approach that will benefit agriculture in developing countries; and

— a source of value-added genes and traits that will increase farmer productivity and profitability.

Status of Plant Biotechnology

Plant Transformation

Remarkable progress has been made in the development and application of gene transfer systems to crops since the first demonstration of transgenic plant production just over ten years ago. Today, over 80 species of crop plants can be genetically manipulated using available Agrobacterium tumefaciens or a variety of free DNA delivery transformation systems (Table 1). This list includes nearly all major dicotyledonous crops and a rapidly increasing number of monocotyledonous crops, including wheat, rice and maize.

It is highly likely that routine gene transfer systems will exist for nearly all crops within the next two or three years. While technical improvements will lead to further increases in transformation efficiency, extend transformation to elite commercial germplasm and lower transgenic plant production cost, there is no significant barrier, even today, to the application of plant transformation to crop improvement.

Gene Expression

Plant genetic engineers currently have in hand a large battery of regulatory sequences that provide for both constitutive expression as well as highly accurate targeting of gene expression to specific tissues within transgenic plants. Moreover, established differential screening methods allow for ready isolation of regulatory sequences that may be required for even more sophisticated expression requirements. The ability to decrease endogenous gene expression in plants represents a remarkably powerful tool, and striking phenotypic alterations have been observed by selective inactivation of genes using antisense technology.

Table 1: Species in which Transgenic Plants have been Produced

Alfalfa	Foxglove	Potato
Apple	Grape	Rapeseed
Arabidopsis	H. albus	Rice
Asparagus	Horse radish	Rose
B. carinata	Kiwi	Rye
B. rapa	Lettuce	S. integrifolium
Brown Sarson	Licorice	Snap Dragon
Cabbage	Lotus	Soybean
Canola	M. truncatula	Strawberry
Carnation	Maize	Sugarbeet
Carrot	Morning glory	Sunflower
Cauliflower	Muskmelon	Sweet potato
Celery	Orchard Grass	Tobacco
Chicory	Orchid	Tomato
Chrysanthemum	Papaya	Tulip
Cotton	Pear	V. officinalis
Cranberry	Peas	Walnut
Cucumber	Petunia	Wheat
Eggplant	Poplar	White spruce
Flax	Poppy	Yam

Achieving even higher levels of gene expression in selected plant organs would increase opportunities for more economic specialty chemical or pharmaceutical production in plants, and site-specific insertion could minimize the variability of gene expression among transformants. However, current expression systems appear sufficient for meeting immediate crop improvement needs.

Methods of Gene Discovery

Advances in methods for the identification and isolation

of new gene coding sequences are of great importance to the engineering of improved plants. The interspecies-specific use of transposons and T-DNA insertion has permitted the tagging and isolation of novel genes from several plant sources. The availability of high resolution physical maps in tomato and Arabidopsis has already led to mapping of several novel loci and new methods will allow direct testing of the isolated DNA for its ability to complement the mutation of interest at each step of the walking process.

Advances in the redesign of coding sequences for plant expression allow for predictable, high-level expression of a variety of nonplant genes in crop plants. Again, while ongoing research efforts will predictably and dramatically increase the probability and efficiency of gene discovery and isolation, it would appear that even with today's methods most genes can be identified and isolated. Even assuming only modest advances in gene discovery methods and progress on the sequence analysis of the Arabidopsis genome in the next ten years, one could infer that gene discovery will not be a limiting element for very long.

Gene Stability

By the end of 1991, nearly five hundred field test experiments evaluating the performance of genetically engineered plants had been carried out in the USA and Europe alone. The overwhelming conclusions from these extensive studies are that newly introduced genes are stable, inherited and are expressed like any other plant gene. This includes a variety of new genes which provide for control of insects, weeds, and plant diseases as well as for quality improvement.

Such traits have already been successfully introduced into several important crop species and genetically engineered soybean, cotton, rice, rapeseed, sugarbeet, tomato, alfalfa, potato and maize crops are expected to enter the marketplace between the years 1995 and 2000. Broad germplasm access will likely require extensive backcrossing or micropropagation efforts. These established methods, although cumbersome, appear adequate for ensuring large germplasm access in most annual and perennial crops.

Plant biology is entering a unique period where both basic research and commercial applications will be limited only by the creativity of the researcher and by funding levels. While there is an obvious need for substantial expansion of our understanding of basic plant biochemistry and physiology in order to fully exploit scientific advances, there are no significant technical hurdles remaining.

The initial wave of research in plant biotechnology has been driven by the seed and agrochemical industries and has appropriately concentrated on the engineering of "agronomic traits" that relate directly to the traditional roles of these industries in farming, such as the control of insects, weeds, and plant diseases. Progress in this area has been exceedingly rapid, and genes conferring these new traits have already been successfully introduced into several important crop species. The status of some of these product candidates is discussed below.

Insect Resistance

The production of plants that naturally control insects has obvious, important implications for crop improvement, and for both the seed and agrochemical industries. Progress in developing insect control in transgenic plants has been initially achieved through the expression in plants of the insect control protein genes of Bacillus thuringiensis (B.t.). B. t. is a naturally-occurring soil bacterium that produces an insect control protein which is lethal to selected insect pests. Most strains of B.t. are toxic to lepidopteran (caterpillar) larvae, although some strains with toxicity to coleopteran (beetle) or dipteran (fly) larvae have also been described. The insect toxicity of B.t. resides in a large protein; this protein has no toxicity to beneficial insects, other animals or humans. The mode-of-action of the B.t. insect control protein involves disruption of K^+ ion transport across brush border membranes of susceptible insects.

Transgenic tomato, tobacco, cotton and maize plants containing the B.t. gene have exhibited tolerance to caterpillar pests in laboratory tests. A novel approach for increasing expression of B.t. genes in plants, which involves restructuring of the DNA coding sequence without altering the encoded amino acid sequence, has led to substantial enhancement in insect control. Cotton plants with a high level of resistance to boll damage by caterpillars have been developed, and commercial levels of control have been achieved with both potato and cotton. Both products are being reviewed by the EPA.

Field tests have confirmed excellent protection (equivalent to weekly insecticide spraying) from bollworm, budworm and pink bollworm. Excellent protection from defoliation by Colorado potato beetle has also been observed in greenhouse and field experiments with potato plants containing the novel coleopteran-active B. t. tenebrionis gene. The insect resistant plants sustained no damage from Colorado beetles throughout the growing season under conditions of high insect pressure.

Other types of insecticidal molecules are clearly necessary to extend biotechnology approaches for controlling additional insect pests in these and other target crops. Extensive efforts are under way to identify other microbial and plant insecticidal proteins. It has been demonstrated that plants genetically engineered to express a proteinase inhibitor gene demonstrate enhanced resistance to a range of insect pests; in vitro studies indicate the alpha-amylase inhibitor protein has broad-spectrum insecticidal activity.

It is highly likely that a large percentage of insect control in annual crops such as cotton, maize and vegetables will be provided by introduced genes in the next 10-20 years. These developments will provide new solutions for insect control and allow for significant reduction in insecticide usage. An important focus of seed companies introducing these new crops will be ensuring that appropriate agronomic and farm management practices are utilised to minimise any possibility of insects developing resistance to the plants.

Weed Control

Engineering tolerance to a specific herbicide into a crop plant represents a new alternative for conferring selectivity and enhancing the crop safety of herbicides. While laboratory experiments have shown that it is possible to achieve resistance to nearly a dozen different herbicides, R&D efforts by private companies have concentrated only on those herbicides with minimal environmental impact, with emphasis on properties such as high unit activity, low toxicity and rapid biodegradation. Care has also been taken to ensure that herbicide-tolerant genes will not be introduced into crops which could become "volunteer" weeds in subsequent crop rotations or which outcross readily with weed species.

The development of crop plants which are tolerant to such herbicides would provide for more effective, less costly and more environmentally attractive weed control options than exist today. The commercial strategy behind engineering herbicide tolerance is to gain market share through a shift in herbicide use, not to increase the overall use of herbicides as is popularly held by critics. Two general approaches have been pursued in engineering herbicide tolerance:

1) altering the level and sensitivity of the target enzyme for the herbicide; and

2) incorporating a gene encoding an enzyme which can inactivate the herbicide.

As an example of the first approach, Roundup® herbicide acts by specifically inhibiting the enzyme 5-enolpyruvylshikimate-3-phosphate synthase. Roundup® herbicide is active against annual and perennial broad leaf and grassy weeds, it has very low animal toxicity, and it is rapidly inactivated and degraded in all soils.

Tolerance to Roundup® herbicide has been engineered into canola, soybean, cotton and maize by introducing genetic constructions for the overproduction of herbicide resistant EPSPS enzymes. Similarly, resistance to sulfonylurea compounds, the active ingredients in Glean®, and Oust® herbicides, has been produced by the introduction of mutant acetolactate synthase (ALS) genes into canola and cotton.

Sulfonylureas are broad spectrum herbicides that are effective at very low application rates. Resistance to gluphosinate, the active ingredient in Basta®, and bromoxynil has been achieved by the alternative approach of introducing bacterial genes encoding enzymes that inactivate the herbicides by acetylation or nitrile hydrolysis respectively. In field tests, gluphosinate-tolerant canola, soybean and maize have shown excellent tolerance to the herbicide. Similarly, bromoxynil tolerant cotton has been extensively evaluated and shown to provide excellent control of broadleaf weeds.

The current crop targets for engineered herbicide tolerance include soybean, cotton, maize, rapeseed and sugarbeet. For the farmer, many factors such as weed spectrum, herbicide performance, environmental impact, seed and chemical cost, application timing and flexibility have to be considered when choosing a particular weed control system. The availability of herbicide tolerance in annual crops over the next decade will give farmers more flexibility in choosing effective and less costly options for weed control.

Herbicide-tolerant plants will have the positive impact of shifting overall herbicide usage through substitution of more effective and environmentally acceptable products. Such improvements in chemical weed control will also allow for higher adoption of minimum tillage practices, and encourage crop rotations which will further reduce soil erosion.

Disease Resistance

Significant resistance to a variety of plant viral diseases has been achieved by coat protein-mediated protection, which involves expressing the coat protein gene of a particular virus in transgenic plants. The mechanism for coat protein-mediated cross protection is likely to involve interference with the uncoating of virus particles in cells prior to translation and replication. Using this approach, results have been obtained for transgenic tomato, alfalfa, tobacco, potato, melon and rice against a broad spectrum of plant viruses, including alfalfa mosaic virus, cucumber mosaic virus, potato virus X (PVX), potato virus Y (PVY) and potato leaf roll virus.

Excellent tolerance has been observed in field tests of Russet Burbank potatoes containing coat protein genes to both PVY and PVX. Recently, very significant resistance to tobacco mosaic virus in tobacco plants has also been obtained by an alternative method which involves expression of a subgenomic viral replicase component.

Rapid progress is also being made in engineering resistance to bacterial and fungal pathogens by several groups around the world. Resistance to the bacterial pathogen Pseudomonas syringae, which causes wildfire in tobacco, has been introduced in transgenic tobacco by expressing a tabtoxin resistance gene that codes for an acetyltransferase. This result demonstrates a successful approach to engineering disease resistance in plants by detoxification of pathogenic toxins.

Some success in engineering resistance to fungal diseases has also been reported. A chitinase gene from the soil bacterium Serratia marcescens was stably expressed in transgenic tobacco. The preliminary results indicated that the expression of the bacterial chitinase in transgenic tobacco leaves resulted in significantly reduced severity of disease caused by a brown-spot pathogen, Alternaria longipes. The plants were reported to have significantly reduced fungal lesions as well as delayed susceptibility to the pathogen.

A bean chitinase gene driven by a high level, constitutive promoter has been expressed in tobacco plants. These plants exhibit increased resistance to the pathogenic fungus Rhizoctonia solani, resulting in significantly reduced root damage and enhanced ability to survive in infested soil. Genes conferring fungal resistance based on the plant's own defense response are being cloned by a number of research groups; one of these proteins, termed osmotin, has been shown to have potent in vitro activity against Phytophthora infestans, the causal agent of late blight disease in potato. Manipulating existing defense mechanisms may also prove to be useful.

Stress Resistance

A variety of abiotic stresses including water, temperature and soil composition are known to impact crop productivity. Although the complexity of plant stress responses has eluded early demonstration of improved phenotypes using plant biotechnology methods, these tools are being applied to dissect and understand the molecular basis for plant response.

A number of plant genes induced by exposure to heat, cold, salt, heavy metals, phytohormones, nitrogen etc., have been identified. Additionally, rapid progress is being made in identifying ion transport pumps and proteins which regulate transport of molecules through channels and plasmodesmata. Metabolites such as proline and betaines have been implicated in stress tolerance in both bacteria and plants-experiments are in progress in a number of laboratories to evaluate the potential of these metabolites to alleviate stress in engineered plants and understand their mode of action. As these advances accelerate, it is highly likely that there will be demonstrations of heat, cold, drought and salt tolerance in the near future.

The world-wide seed and agrochemical industries have been the leading commercial sponsors of agricultural biotechnology research. However, the technical advances and the success of the first products has increased interest by the food processing, specialty chemical and pharmaceutical industries (Figure 1).

Food Processing

Plant biotechnology offers exciting opportunities for the food processing industry to develop new and more nutritious food products and cost effective processes. Examples of such applications include production of higher quantities of sugars, starch or specialised starches

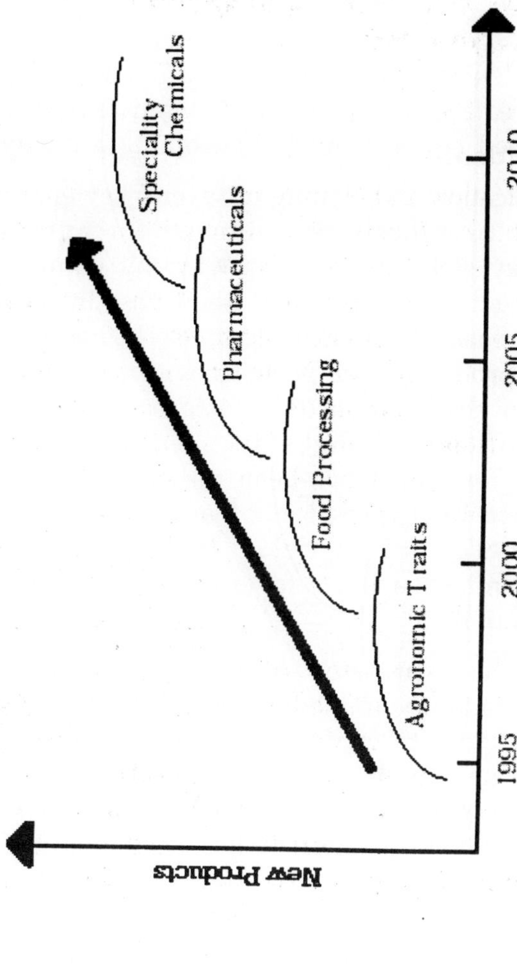

Figure 1. Plant Biotechnology Promises to Deliver Many New Products in Coming Decades

with various degrees of branching and chain length to improve texture, storage and cooking properties.

Starch levels have been increased by 20-40% in potato by expression in tubers of a bacterial ADP glucose pyrophospho-rylase gene (ADPGPPase), which has been shown to be the rate-limiting step in starch production in plants. Expression of the gene encoding sucrose phosphate synthase in transgenic tomato plants has been shown to elevate sucrose levels and reduce starch. The opportunity to specifically create specialty oils or to eliminate particular fatty acids in seed crops is quickly becoming a reality.

The identification and cloning of several key enzymes in fatty acid biosynthesis has allowed for antisense inhibition of key steps in the pathway; by expressing antisense constructs to stearyl ACP desaturase in canola, significant increases in stearic acid levels have been observed. The production of proteins with nutritionally balanced amino acid compositions has been shown to result in significant changes in seed amino acid composition. In tomato, it has been possible to increase bruise resistance by expression of antisense RNA to polygalacturonase and to delay fruit ripening by expression of antisense RNA to enzymes involved in ethylene production.

Recently, the expression of ACC deaminase, an enzyme that degrades the immediate precursor to ethylene, has also been shown to confer controlled ripening. The enzymes and genes involved in biosynthesis of coloring materials and flavors are also important to the food industry and to the consumer; it has been possible to manipulate color of flowers by sense and antisense expression of the flavonoid biosynthetic genes.

Specialty Chemical Products

Enormous opportunity lies in the successful exploitation

of crops for both commodity and specialty chemical products. Plants have traditionally been a source of a wide range of monomeric and polymeric materials. These range from sugars and fatty acids to polymers such as starch and celluloses, which are carbohydrate based, and to polyhydrocarbons such as rubber and waxes. Many of these polymers have been replaced in the last two to three decades by synthetic materials derived from petroleum-based products.

However, the cost, supply and waste-stream problems often associated with petroleum-based products are issues which have focused new attention on the use of renewable, biological materials. Genetic engineering will significantly enlarge the spectrum and composition of available plant monomers and polymers. Expression of an Escherichia coli mannitol dehydrogenase gene in tobacco has allowed for increased levels of mannitol in plants.

Cyclodextrins are interesting specialty starches which have application in catalysis, formulations, and food processing; introduction of a bacterial cyclodextrin glucosyl transferase has been reported to result in low but detectable levels of cyclodextrins in potato tubers. Recently, researchers have reported the expression of acetoacetyl CoA synthetase and the acetoacetyl CoA reductase in transgenic plants-these two enzymes constitute the two steps leading to the production of poly-hydroxybutyrate, an interesting thermoplastic polymer. It has been proposed that expression of a novel fatty esterase could result in production of C-12 fatty acids in temperate crops like soybean and canola.

Pharmaceuticals

Plants also offer the potential for production of foreign proteins with a variety of health-care applications. Proteins such as neuropeptides, blood factors and growth hormones could be produced in plant seeds and this may ultimately prove to be an attractive economical means of production. Several mammalian proteins have already been produced in genetically engineered plants, including pharmaceutically-active peptides, enkephalins, in rapeseed and human serum albumin in potato.

Plant virus vectors containing desired genes may represent a particularly interesting route for producing large quantities of proteins in plants, due to their high copy number. In the longer term, there is little doubt that biotechnology will be used to improve the nature of both the micro- and macro-ingredients in plant-derived foods. The relationship between diet and disease is slowly emerging, and the consumer demand for more healthy foods is expected to grow.

From the above examples, it is clear that plant biotechnology will have an enormous impact on the seed, agrochemical, food processing, specialty chemical and pharmaceutical industries of the industrialised world. However, the majority of the five billion additional people that will live on our planet in 2030 will be in the less developed or developing countries. Most experts agree that biotechnology will, and in fact must, have a positive impact on agriculture in the developing world. It will not be a "quick fix" to the problems of food production and world hunger, and it will not substitute for conventional agricultural applications, economic and political reform, education, solutions to rural landlessness, international debt relief or population control, among others.

But biotechnology can make big contributions in helping to ensure a sustainable supply of adequate food. One of the key reasons plant biotechnology will have this impact is because of its inherently low capital cost for delivery and implementation. Biotechnology enables delivery of the latest technology to a rural farmer in the package he is most familiar with-the seed. In fact, "plant biotechnology" has been identified by the World Bank, the United Nations Educational, Scientific and Cultural Organisation (UNESCO), the French Institute for Scientific Research for Development and Cooperation (ORSTOM), the United States Agency for International Development (USAID), the Rockefeller Foundation, the International Service for the Acquisition of Agri-Biotech Applications (ISAAA) and numerous other international aid and scientific groups as a high priority in order to meet the future agricultural production needs of developing countries.

Because of the unique challenges associated with regional population density, food distribution, and dietary preference, agricultural biotechnology must be developed locally in developing countries. Some of the issues which currently must be addressed to facilitate biotechnology in developing countries are:

— identification of near-term crop/trait targets; . mobilisation of resources and funding;

— establishment of technology transfer mechanisms with private/public organisations;

— development of a regulatory oversight framework; and

— addressing of property rights and trade issues.

Success will clearly require a coordinated effort between international agencies and governments as well as private companies and public institutions. Several private companies have taken a leadership role in technology transfer to the developing world by establishing relationships which bring these important constituent groups together. Monsanto Company, for example, has projects with Mexico and Africa, which are sponsored by the Rockefeller Foundation and USAID respectively, to develop virus resistant potatoes and other root crops. The benefits of such interactions are both short and long term. Obvious and most important, in the short term, is the opportunity to help alleviate suffering and poverty; in the long term, it is hoped these countries will become important trading partners.

Commercialisation Barriers

Despite the phenomenal scientific progress that has underscored the development of the first generation of plant biotechnology products, significant and serious challenges remain which could delay or even prevent their successful introduction and acceptance in the marketplace.

Regulatory Approval

Field testing and commercial introduction of genetically engineered products come under the statutory jurisdiction of three federal agencies in the USA: the United States Department of Agriculture (USDA), the Food and Drug Administration (FDA) and the Environmental Protection Agency (EPA). In Europe, regulation is based on individual country guidelines, but efforts are progressing to develop a policy on field release to cover the entire European Union (formerly the European Community).

The field testing of genetically engineered crops has proceeded remarkably smoothly; since 1987 there have been thousands of tests of engineered crops in diverse locations across the USA and Europe. These field tests, along with thousands of laboratory experiments, have produced a remarkable degree of consensus internationally that engineered crops present virtually no risk to the environment or to human or animal health. The regulatory emphasis has now shifted from laboratory and field testing to commercialisation approval.

In the USA, progress has been made to develop a coordinated framework between USDA, FDA and EPA for oversight of the commercialisation of engineered crops and food products. Timing is particularly critical since the defining of regulatory approval requirements sets both projected regulatory costs and commercialisation timelines for new products; this is a significant issue for the universities, institutes and companies currently in the process of developing improved genetically engineered crops for introduction in the mid-1990s.

It is essential that regulatory requirements for transgenic plants continue to be based on scientific principles that protect the safety of consumers and the food supply, and recognise the inherent low risk of gene transfer technology and the benefits afforded by genetically engineered crops to growers, food processors and consumers. International harmonisation of regulatory approval standards is necessary to prevent development of non-tariff trade barriers between countries that could ban shipments of genetically engineered seeds or food products without sound scientific basis. Other imposed political barriers could include labeling and inspection "requirements" that would effectively limit access and distribution of these products.

Patent Protection

Patent protection for genetically-engineered plants is essential to offset the cost of developing crops with significant new traits, and to encourage more overall investment and research in this field. Patent protection provides a broader proprietary right than is currently provided under either the International Convention for the Protection of New Varieties of Plants (UPOV) or the Plant Variety Protection Acts (PVPA) in the USA.

While few dispute that companies that have invested heavily in R&D to isolate, test and commercialise genes are entitled to protection for their inventions, there is considerable debate within the seed industry concerning how much protection is deserved and how patents will impact the cooperative nature of the seed industry itself. Much of this concern results from confusion surrounding the restrictions imposed by patent rights versus the incentive they provide for competitive research and product development which stimulates innovation.

Many of the conciliatory proposals, including patenting of genes (but not plants) as well as compulsory licensing in the event that plant patenting is permitted, if implemented, could significantly reduce the incentive for private industry funding. Lack of proprietary protection for genetically engineered plants in many countries outside the USA remains a very serious limitation; plant and animal varieties are largely excluded from patent protection by the European countries that signed the 1973 European Patent Convention. The availability, scope and enforceability of plant patents in other countries, including Japan, China and Eastern European countries, remains questionable; perhaps more than any other issue, proprietary rights protection may limit plant biotechnology R&D and commercialisation in these regions.

In today's era of global communications, radical groups opposed to technological creativity have demonstrated an increasing ability to influence public and political opinion. As opposed to the more mainstream publicinterest groups with whom dialogue is possible, these groups exploit the public's lack of understanding of biotechnology and interweave political, societal and emotional issues (population control, animal rights, religion, concerns over the environment, etc.) for purposes of gaining visibility, credibility, membership, financial support and political presence.

These critics represent various positions across a spectrum of thought; in most cases, they rarely attack science and technology directly, but instead focus on peripheral, but more vulnerable issues, such as food safety, proprietary rights, religious views, etc. Popular strategies involve reducing the commercial incentive for R&D through actions which delay product introduction and increase commercialisation costs. For example, proposals for mandatory licensing of patented plants are often supported because this would reduce the incentive for companies to invest in the development of proprietary new crops.

By attempting to block critical and needed components of plant biotechnology such as the use of marker genes in plants through political, nonscience-based regulation, some of these groups hope to effectively delay commercialisation or create opportunities for eventual trade sanctions. Similarly, by falsely raising concerns over food safety, they hope to increase regulatory testing costs to prohibitive levels. Others are demanding labeling of foods to discourage consumer purchase and also to facilitate subsequent boycotting activities against particular brands or companies who introduce, distribute or sell these products. These groups represent a serious threat; they are just as motivated, and often as well-funded, as the companies and organisations who are trying to commercialise plant biotechnology products. Their misrepresentations of issues and concerns are dangerously effective at further confusing a skeptical public and undermining the science-based regulation that is in place to protect the food system.

4

Modern Biotechnology for Food and Agriculture

In developing countries, there is very little commercial utilisation of results from modern biotechnology research. As a result, the potential contributions of biotechnology to poverty alleviation and enhanced food security and nutrition in developing countries has received little attention, beyond blanket statements of support or opposition. Small-scale farmers in developing countries are faced with many problems and constraints. Preand postharvest crop losses due to insects, diseases, weeds, and droughts result in low and fluctuating yields, as well as risks and fluctuations in incomes and food availability.

Low soil fertility and lack of access to reasonably priced plant nutrients, along with acid, salinated, and waterlogged soils and other abiotic factors, contribute to low yields, production risks, and degradation of natural resources as poor farmers try to eke out a living. They are often forced to clear forest or farm ever more marginal land to cultivate crops. Poor infrastructure and poorly functioning markets for inputs and outputs together with lack of access to credit and technical assistance add to the impediments facing these farmers. These farmers and other rural and urban poor people suffer from food insecurity and poor nutrition, caused in large measure by poverty and lack of nutritional balance in the diet they can afford.

About 1.2 billion people, or one of every five humans, live in a state of absolute poverty, on the equivalent of US$1/day or less. About 800 million people are food insecure, and 160 million preschool children suffer from energy-protein malnutrition, which results in the death of over 5 million children under the age of five each year. A much larger number of people suffer from deficiencies of micronutrients such as iron and vitamin A. For example, 2 billion people (one of every three) are anemic, usually as a result of iron deficiency. Food insecurity and malnutrition result in serious public health problems and lost human potential in developing countries.

Around 70 percent of poor and food-insecure people reside in rural areas, although poverty and food insecurity appear to be growing in urban areas as urbanisation proceeds apace in developing countries. The World Bank forecasts that poverty's center of gravity will remain rural in the early decades of the 21st century. Most rural poor people depend directly or indirectly on agriculture for their livelihood. Poor people in rural or urban areas spend as much as 50-70 percent of their incomes on food .

Low productivity in agriculture is a major cause of poverty, food insecurity, and poor nutrition in low-income developing countries. This is true for urban and rural poor people alike. Low productivity means low incomes for farmers and farm workers, little demand for goods and services produced by poor nonagricultural households in the rural areas, and unemployment and underemployment in urban areas. It also means high unit costs for food, which translate into reduced consumer purchasing power. High food prices are a serious matter for households that spend a large share of their budget on food.

In low-income developing countries, agriculture is the driving force for broad-based economic growth and poverty alleviation. A healthy agricultural economy offers farmers incentives for sound management of the natural resource base upon which their livelihood depends. These relationships are borne out not only by research but also by history in both developing and industrial nations. Productivity increases in European and U.S. agriculture were extremely important to broad-based economic growth during earlier periods of development.

More recently, productivity increases in agriculture, led by agricultural research - the Green Revolution -formed the locomotive of rapid broad-based economic growth and poverty reduction in many Asian countries, including China, Indonesia, South Korea, and India. Recent IFPRI research in four African countries found similar strong linkages between agricultural productivity growth and general economic growth (Delgado and others 1998).

Productivity gains are essential not only for economic growth and poverty alleviation, but to assure that food supplies remain adequate for a growing world population. According to United Nations projections, world population will increase by 25 percent to 7.5 billion in 2020. On average, 73 million people will be added annually. Over 97 percent of the projected growth will take place in developing countries.

Poverty Alleviaiton through Agriculture

Agriculture must figure prominently in poverty alleviation strategies of developing countries. Accelerated public investments are needed to facilitate agricultural and rural growth through:

— Yield-increasing crop varieties, including those that are drought and salt tolerant and pest resistant, and improved livestock

— Yield-increasing and environmentally friendly production technology

— Reliable, timely, and reasonably priced access to appropriate inputs such as tools, fertiliser, and, when needed, pesticides, as well as the credit often needed to purchase them

— Strong extension services and technical assistance to communicate timely information and developments in technology and sustainable resource management to farmers and to relay farmer concerns to researchers

— Improved rural infrastructure and effective markets

— Particular attention to the needs of women farmers, who grow much of the locally produced food in many developing countries

— Primary education and health care, clean water, safe sanitation, and good nutrition for all.

These investments need to be supported by good governance and an enabling policy environment, including trade, macroeconomic, and sectoral policies that do not discriminate against agriculture, and policies that provide appropri-ate incentives for the sustainable management of natural resources, such as secure property rights for small farmers. Development efforts must engage poor farmers and other low-income people as active participants, not passive recipients; unless the affected people have a sense of ownership, development schemes have little likelihood of success.

Developing countries must reverse present declining levels of public investment in agriculture. On average, they devote 7.5 percent of government spending to agriculture (and just 7 percent in Sub-Saharan Africa). For their part, donor countries must redress the precipitous decline in aid to agriculture and rural development, which plunged by nearly 50 percent in real terms between 1986 and 1996. Overall development aid has also fallen in recent years. Donors must also rethink their rather inflexible emphasis of the past two decades on less government and a smaller public sector, which has contributed to public disinvestment in agriculture in the developing countries.

Investment in Agricultural Research

Public investment in agricultural research is of particular importance for achieving food security in developing countries. The private sector is unlikely to undertake much of the research needed by small farmers because it cannot expect sufficient returns to cover costs. IFPRI research has shown that the annual rates of return to agricultural research and development are, on average, 73 percent.

Benefits to society from agricultural research can be extremely large but will not be obtained without public investments. Low-income developing countries grossly underinvest in agricultural research: less than 0.5 percent of the value of their agricultural production, compared to 2 percent in higher-income countries. Sub-Saharan Africa, which desperately needs productivity increases in agriculture, has only 42 agricultural researchers per million economically active persons in agriculture, compared with 2,458 in industrial countries.

Efforts to improve longer-term productivity on small-scale farms, with an emphasis on staple food crops, must be accelerated. Research and policies are also needed to help farmers, communities, and governments better cope with risks resulting from such factors as poor market integration, poorly functioning markets, and climatic fluctuations. More research must be directed to the development of appropriate technology for sustainable intensification of agriculture in resource-poor areas, where a high percentage of poor people live, and where environmental risks are severe. The needed research must join all appropriate scientific tools together, with better use of the insights of traditional indigenous knowledge.

Research and technology alone will not drive agricultural growth. The full and beneficial effects of agricultural research and technological change will materialise only if government policies are conducive to and supportive of poverty alleviation and sustainable management of natural resources.

Agricultural Biotechnology and Food Security

Although conventional applications of biotechnology, such as tissue culture and fermentation amongst others, is under way in several developing countries, little genetically improved (transgenic) seed material has been grown in the poorer developing countries to date so ex post assessment is virtually impossible. A great deal is known, however, about the social and economic risks and benefits associated with traditional Mendelian plant breeding as exemplified by the Green Revolution.

The analysis, therefore, begins with the identification of similarities and differences between the Green Revolution and modern biotechnology, and an attempt is made to draw lessons from the Green Revolution and to look at the difference between that technology package and modern biotechnology to try to assess the likely social and economic risks and benefits of modern agricultural biotechnology ex ante.

Green Revolution and Modern Biotechnology

There are three differences of particular importance for an assessment of social and economic risks and benefits. The research leading to the Green Revolution was undertaken by the public sector and the improved seed was usually freely available for seed multiplication and distribution.

Although breeders' rights may permit an initial charge for the improved materials, the intellectual property rights (IPR) did not extend beyond the initial release. Having acquired the seed, farmers could reuse it without further payment, although reuse of hybrid seed would drastically reduce the yield advantage.

This is in keeping with the principle of "farmers' rights" included in the 1983. International Undertaking on Plant Genetic Resources. In contrast, the bulk of modern agricultural biotechnology research is undertaken by private sector firms, which protect IPRs through patents that extend beyond the first release. Farmers, therefore, cannot legally plant or sell for planting the crop produced with the patented seed without the permission of the patent holder.

Patent holders, currently seeking ways to enforce their rights, are considering approaches such as legal agreements and technologies that will activate and deactivate specific genes. However, monitoring and enforcing contracts that prohibit large numbers of small farmers from using the crops they produce as seed would be expensive and difficult.

The so-called terminator gene is the first patented technology aimed at biological IPR protection. It is not appropriate for small farmers in developing countries because existing infrastructure and production processes may not be able to keep fertile and infertile seeds apart. Small farmers could face severe consequences if they planted infertile seeds by mistake. Commercialisation of the terminator gene now seems unlikely in the short term.

Research is under way on other biological approaches to IPR protection that would not impose such risk on small farmers. These include, for example, genetically engineered seeds that contain desired traits, such as pest resistance or drought tolerance, but in which these are activated only through chemical treatment. Otherwise, the seed would maintain its normal characteristics. Thus, if a farmer planted an improved seed, the offspring would not be sterile; rather they would revert to normal seeds, without the improved traits. The farmer would have the choice of planting the seed and doing no more, or activating the improved traits by applying the chemical. This approach complies with the principle of doing no harm.

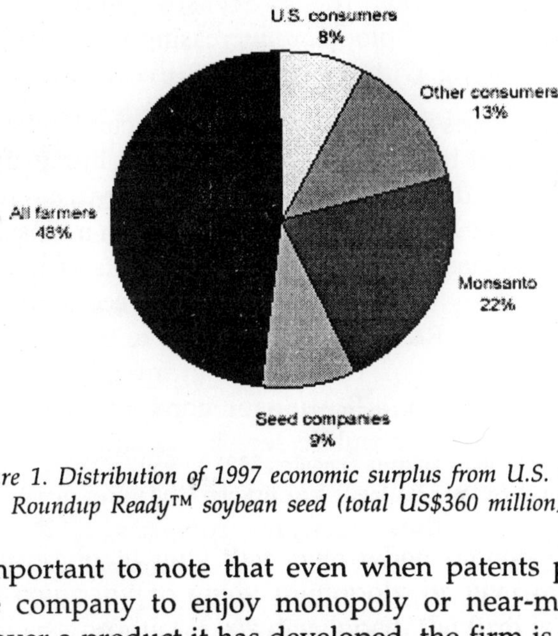

Figure 1. Distribution of 1997 economic surplus from U.S. use of Roundup Ready™ soybean seed (total US$360 million)

It is important to note that even when patents permit a private company to enjoy monopoly or near-monopoly rights over a product it has developed, the firm is unlikely to capture 100 percent of the economic benefits. A recent study of the distribution of the economic benefits generated by the use of herbicide-tolerant soybean seed in the United States in 1997 found that the company, Monsanto, received 22 percent, while seed companies gained 9 percent. Consumers of soybean and soybean products in the United States and other countries reaped a 21 percent share, whereas farmers worldwide obtained 48 percent (Figure 1). The share of U.S. farmers was actually 51 percent of the benefits, but farmers elsewhere experienced net losses of 3 percent.

A second, and related, difference between the Green and Gene revolutions involves the patenting of processes as well as products. The main process behind the Green Revolution was conventional plant breeding technology, which lies in the public domain, carried out by public institutions. Today, the processes used in modern agricultural biotechnology are increasingly subjected to IPR protection, along with the products that result.

This means that public sector research institutions may not be able to gain access to basic but proprietary knowledge and processes needed in research, including research on the so-called orphan crops such as cassava and millet. These are critical staples in the diets of many poor people, but they do not offer promising economic returns to private sector R&D efforts, so efforts to develop disease-resistant cassava or drought-tolerant millet, whether through genetic modification or conventional breeding, must come from the public sector.

Some firms have agreed to transfer proprietary technologies, without charging royalties, to developing countries where there are few potential commercial prospects. Monsanto, for example, has entered into agreements with Kenyan and Mexican government agricultural research institutes to develop virus-resistant crops. Arrangements such as these are few and generally involve the philanthropic arms of the private firms.

A third difference involves the adaptation of industrial country agricultural research to developing country conditions. Although based on earlier research in industrial countries, the Green Revolution was focused on solving specific problems in developing countries. Current application of modern biotechnology is focused on industrial country agriculture.

Industrial country research institutions had begun working on development of higher yielding crop varieties in the late 19th century. For example, in Japan, rice breeding under the auspices of the Ministry of Agriculture and public universities led to large yield gains in the early part of the 20th century, with a second wave of major gains after 1945.

During the early decades of Soviet history, under the leadership of geneticist Nikolai Ivanovich Vavilov, the government carried out extensive crop improvement programmes and established one of the world's largest germplasm collections. In the United States, hybrid maize research began in the 1920s. Much of the basic research was done by public institutions, such as land grant universities, state experiment stations, and the U.S. Department of Agriculture (USDA). Applications to particular farming conditions and the mass marketing of the new varieties were, in turn, handled by private seed firms such as Pioneer Hi-Bred and DeKalb. The research focused not only on developing higher yielding seeds to bolster food supplies for domestic consumption which was a critical U.S. concern up to the 1940s), but also on animal feed and production for export. This research could not simply be transferred to poorer developing countries, where the need was for improved varieties of locally-consumed staples.

The research that led to the Green Revolution involved further adaptation to the agroecological conditions of tropical and semitropical areas. It also focused on rice, wheat, maize, root and tuber crops, and tropical fruits and vegetables. The public sector role was, if anything, even more prominent, with international agricultural research centers (IARCs) and national agricultural research systems (NARS),particularly in Asia and Latin America, playing a prominent role.

Financial support came from donors of official development assistance and large private foundations, such as Ford, Rockefeller, and Kellogg. In contrast, modern agricultural biotechnology is still in an early phase, and the focus is overwhelmingly on production on industrial country farms and for industrial country markets. In 1998, 85 percent of the land planted to genetically improved (GI) crops was in just five developed countries (Australia, Canada, France, Spain, and the United States), with the United States alone accounting for about 75 percent of the area. Argentina, China, Mexico, and South Africa cultivated the remaining 15 percent, and the countries other than China include a substantial number of large-scale, capital-intensive farms that produce primarily for industrial country markets. Among the crops produced in these four developing countries are insect-resistant cotton and maize, herbicide-resistant soybean, and tomatoes with a long shelf life.

Globally, herbicide-resistant soybean, insect-resistant maize, and genetically improved cotton (containing insect resistance and/or herbicide tolerance genes) account for 85 percent of all plantings. Both the area planted to genetically improved crops and the value of the harvests grew dramatically between 1995 and 1999: from less than 1 million hectares to 28 million in 1998 and approximately 40 million in 1999, and from US$75 million in 1995 to US$1.64 billion in 1998. Private industry has dominated research (there are a few exceptions: for example, Rockefeller Foundation support for research on rice, USDA's role in developing the terminator technology, and modest programmes at IARCs). Consolidation of the industry has proceeded rapidly since 1996, with more than 25 major acquisitions and alliances worth US$15 billion.

Little private-sector agricultural biotechnology research so far has focused on developing country food crops other than maize. Moreover, little adaptation of the research to developing country crops and conditions has occurred through the "enlightened" (that is, not for profit, public goods oriented) public and philanthropic channels prominent in the Green Revolution of the developing countries. A programme directed at public/private sector linkages is that of the International Service for the Acquisition of Agri-biotech Applications (ISAAA), which transfers and delivers appropriate biotechnology applications to developing countries and builds partnerships amongst institutions.

Relatively little biotechnology research currently focuses on the productivity and nutrition of poor people. As with the Green Revolution, the challenge is to move from the scientific foundation established by industrial country-oriented research efforts to research focused on the needs of poor farmers and consumers in developing countries. Direct transfers of the fruits of agricultural biotechnology research to the developing countries will not work, in most cases.

More appropriate research for the developing world might focus on biotechnology and conventional breeding to develop alternative forms of weed resistance, such as leafier rice that denies weeds sunlight rather than incorporating herbicide tolerance into rice. The West Africa Rice Development Association (WARDA), a public IARC in Côte d'Ivoire, has used a combination of conventional plant breeding and tissue culture to develop such rice.

Insect-resistant crops would have great potential value for poor farmers. So far, however, the development of crops containing genes from the Bacillus thuringiensis (Bt) bacterium, which produces a natural pesticide, has focused largely on the crops and cropping environments of North America. The new crop varieties containing the Bt gene require extremely knowledge-intensive cultivation. They might well be transferable to larger scale operations in some developing countries such as Argentina.

The potential usefulness of this application in crops grown by small farmers is open to question. There is considerable debate about risks of the development of resistance in pests, harm to beneficial insects, and crosspollination of wild and weedy plants with the novel gene. The evidence on these issues is still inconclusive and warrants careful monitoring before the application of Bt is tried on a large scale in crops grown by subsistence farmers.

Research on crops and problems of relevance to small farmers in developing countries will require the allocation of additional public resources to agricultural research, including biotechnology research, that promises large social benefits. There is no reason to believe that this research will offer lower rates of return than other agricultural research and development.

Private-sector agricultural research currently accounts for a small share of agricultural research in most developing countries. The public sector can expand private-sector research for poor people by converting some of the social benefits to private gains, for example, by offering to buy exclusive rights to newly developed technology and make it available either for free or for a nominal charge to small farmers. The private research agency would bear the risks, as it does when developing technology for the market. IARCs have an important role to play as intermediaries in facilitating such arrangements.

Without more enlightened adaptation, continued expansion of genetically improved crop production in the industrial countries may well have a negative impact on small farmers in developing countries. Some developing country consumers would benefit, but those consumers who also farm could experience net losses. In addition, the development of industrial substitutes for developing country export crops, such as cocoa (which in many developing countries is produced by small farmers) could have a devastating impact on developing country farmers' livelihoods. In sum, the biggest risk of modern biotechnology for developing countries is that technological development will bypass poor farmers and poor consumers because of a lack of enlightened adaptation.

It is not that biotechnology is irrelevant, but that research needs to focus on the problems of small farmers and poor consumers in developing countries. Private sector research is unlikely to take on such a focus, given the lack of future profits. Without a stronger public sector role, a form of "scientific apartheid" may well develop, in which cutting edge science becomes oriented exclusively toward industrial countries and large-scale farming.

The outcomes of the Green Revolution offer some guideposts for assessing the likely risks and benefits of agricultural biotechnology for developing countries. Risks and benefits may be inherent in a given technology, or they may transcend the technology. The policy environment into which a technology is introduced is critical. For example, IFPRI research has found that in Tamil Nadu State in India, the adoption of high-yielding grain varieties meant not only increased yields and cheaper, more abundant food for consumers, but income gains for small and largerscale farmers alike, as well as for nonfarm poor rural households.

Increased rural incomes contributed to nutrition gains for these households. Because the Tamil Nadu state government has pursued active poverty alleviation strategies, including extensive social safety net programmes and investment in agriculture, rural development, and a fair measure of equity in access to resources such as land and credit, the benefits were widely shared.

Where increased inequality followed the adoption of Green Revolution technology, it was not because of factors inherent to the technology, but rather a result of policies that did not promote equitable access to resources. And even in these areas, rural landless laborers usually found new job opportunities as a consequence of increased agricultural productivity, particularly where appropriate physical infrastructure and markets developed.

Successful adoption of Green Revolution technology, however, depended on access to water, fertilizer, and pesticides. Thus, inequality between well-endowed and resource-poor areas increased because of the properties of the technology itself. Likewise, excessive or improper use of chemical inputs led to adverse environmental impacts in some instances. This problem was offset, to some extent, by characteristics that were also inherent in the technology: by allowing yield gains without expanding cultivated area, the technology kept cultivators from clearing forests and moving onto wild and marginal lands.

Overall, the Green Revolution was extremely successful in enhancing productivity in rice, wheat and maize; in increasing incomes and reducing poverty; and in preserving forests and marginal lands by improving yields within existing cultivated areas. By reducing unit costs and prices for food, it greatly benefited poor consumers, and by boosting farmers' incomes, it contributed to gains in nutrition. Would agricultural biotechnology produce similar results in developing countries? The answer depends on whether the research is relevant to poor people and on its ownership, that is, the nature of the intellectual property rights arrangements.

Risks and Benefits

Modern biotechnology is not a silver bullet for achieving food security, but, used in conjunction with traditional or conventional agricultural research methods, it may be a powerful tool in the fight against poverty that should be made available to poor farmers and consumers. It has the potential to help enhance agricultural productivity in developing countries in a way that further reduces poverty, improves food security and nutrition, and promotes sustainable use of natural resources. Solutions to the problems facing small farmers in developing countries will benefit both farmers and consumers.

The benefits of new genetically improved food to consumers are likely to vary according to how they earn their income and how much of their income they spend on food. Consumers outnumber farmers by a factor of more than 20 in the European Union, and Europeans spend only a tiny fraction of their incomes on food. Similarly, in the United States, farms account for less than 2 percent of all households, and the average consumer spends less than 12 percent of income on food.

In the industrial countries, consumers can afford to pay more for food, increase subsidies to agriculture, and give up opportunities for better-tasting and better-looking food. In developing countries, poor consumers depend heavily on agriculture for their livelihoods and spend the bulk of their income on food. Strong opposition to GI foods in the European Union has resulted in restrictions on modern agricultural biotechnology in some countries. The opposition is driven in part by perceived lack of consumer benefits, uncertainty about possible negative health and environmental effects, widespread perception that a few large corporations will be the primary beneficiaries, and ethical concerns.

Potential Benefits

There are many potential benefits for poor people in developing countries. Biotechnology may help achieve the productivity gains needed to feed a growing global population, introduce resistance to pests and diseases without costly purchased inputs, heighten crops' tolerance to adverse weather and soil conditions, improve the nutritional value of some foods, and enhance the durability of products during harvesting or shipping. New crop varieties and biocontrol agents may reduce reliance on pesticides, thereby reducing farmers' crop protection costs and benefiting both the environment and public health.

Biotechnology research could aid the development of drought-tolerant maize and insect-resistant cassava, to the benefit of small farmers and poor consumers. Research on genetic modification to achieve appropriate weed control can increase farm incomes and reduce the time women farmers spend weeding, allowing more time for the child care that is essential for good nutrition.

Biotechnology may offer cost-effective solutions to micronutrient malnutrition, such as vitamin A- and iron-rich crops. Research focused on how to reduce the need for inputs and increase the efficiency of input use could lead to the development of crops that use water more efficiently and extract phosphate from the soil more effectively. The development of cereal plants capable of capturing nitrogen from the air could contribute greatly to plant nutrition, helping poor farmers who often cannot afford fertilisers.

By raising productivity in food production, agricultural biotechnology could help further reduce the need to cultivate new lands and help conserve biodiversity and protect fragile ecosystems. Productivity gains could have the same poverty-reducing impact as those of the Green Revolution if the appropriate policies are in place. Policies must expand and guide research and technology development to solve problems of importance to poor people. Research should focus on crops relevant to small farmers and poor consumers in developing countries, such as ba-nana, cassava, yam, sweet potato, rice, maize, wheat, and millet, along with livestock.

Health and environmental risks. Genetically improved (GI) foods are not intrinsically good or bad for human health. Their health effect depends on their specific content. GI foods with a higher iron content are likely to benefit iron-deficient consumers. But the transfer of genes from one species to another may also transfer characteristics that cause allergic reactions. Thus, GI foods need to be tested for allergy transfers before they are commercialised. Such testing avoided the possible commercialisation of soybeans with a Brazil nut gene.

GI foods with possible allergy risks should be fully labeled. Labeling may also be needed to identify content for cultural and religious reasons or simply because consumers want to know what their food contains and how it was produced. While the public sector must design and enforce safety standards as well as any labeling required to protect the public from health risks, other labeling might best be left to the private sector in accordance with consumer demands for knowledge.

Failure to remove antibiotic-resistant marker genes used in research before a GI food is commercialised presents a potential although unproven health risk. Recent legislation in the European Union requires that these genes be removed before a GI food is deemed safe.

Risks and opportunities associated with GI foods should be integrated into the general food safety regulations of a country. International agencies and donors may need to assist some developing countries build the capacity to develop appropriate regulatory arrangements. These regulatory systems are needed to govern food safety and assess any environmental risks, monitor compliance, and enforce such regulations. The regulatory arrangements should be country-specific and reflect relevant risk factors.

Progress on achieving a global agreement on biosafety standards is urgently needed. The development of a public global regulatory capacity has lagged far behind the pace of economic globalisation. The ecological risks policymakers and regulators need to assess include the potential for spread of traits such as herbicide resistance from genetically improved plants to unmodified plants (including weeds), the buildup of resistance in insect populations, and the potential threat to biodiversity posed by widespread monoculture of genetically improved crops.

Seeds that allow farmers the option of "turning off" genetic characteristics, mentioned earlier, offer great promise for assuring that new traits do not spread through cross-pollination. Both food safety and biosafety regulations should reflect international agreements and a given society's acceptable risk levels, including the risks associated with not using biotechnology to achieve desired goals. Poor people should be included directly in the debate and decisionmaking about technological change, the risks of that change, and the consequences of no change or alternative kinds of change.

Socioeconomic risks. Unless developing countries have policies in place to ensure that small farmers have access to delivery systems, extension services, productive resources, markets, and infrastructure, there is considerable risk that the introduction of agricultural biotechnology could lead to increased inequality of income and wealth. In such a case, larger farmers are likely to capture most of the benefits through early adoption of the technology, expanded production, and reduced unit costs.

Growing concentration among companies engaged in agricultural biotechnology research may lead to reduced competition, monopoly or oligopoly profits, exploitation of small farmers and consumers, and extraction of special favors from governments. Effective antitrust legislation and enforcement institutions are needed, particularly in small developing countries where one or only a few seed companies operate. Global standards regarding industrial concentration must also be developed; international public policies in this area have not kept pace with economic globalisation.

Effective legislation is also required to enforce IPRs, including those of farmers to germplasm, along the lines agreed to within the WTO and the Convention on Biological Diversity. Ethical questions. A major ethical concern is that genetic engineering and "life patents" accelerate the reduction of plants, animals, and microorgan-isms to mere commercial commodities, bereft of any sacred character. This is far from a trivial consideration. However, all agricultural activities constitute human intervention into natural systems and processes, and all efforts to improve crops and livestock involve a degree of genetic manipulation. Continued human survival depends on precisely such interventions.

5

Genetically Modified Foods

Consumers' attitudes toward biotechnology are divided. A recent study, showed that 31% favor biotechnology, 18% oppose it, 26% have mixed feelings, and 26% neither know nor care. Even within each of these groups, their opinions are as diverse as the people expressing them. The differential acceptance of genetically modified products among consumers can be attributed to the different ways in which they process information about biotechnology and related products. Some people carefully weigh potential benefits more heavily than risks. Others form their biotechnology attitudes solely based on "sound bites" they hear on TV or at work.

Regardless of what attitude a person has toward biotechnology, there are two general ways or routes in which they are persuaded and their attitudes formed: A central route to persuasion and a peripheral route to persuasion. When a person is motivated to understand an issue, when they have the ability to understand the issue, and when they have the opportunity to do so, their attitude will be formed through a central route of attitude formation.

If, however, a consumer is not motivated to understand the issue, or if they do not have the technical or cognitive ability, or the opportunity to think about it, any message they hear will be peripherally processed. It is important to understand how central and peripheral processing influence attitudes before assessing the effectiveness of marketing and education efforts. Figure 1 illustrates the two different routes of forming attitudes towards biotechnology and some of the factors that influence the resulting attitudes along the two paths.

Recent studies have shown that consumers generally view genetic engineering technology as a risky process. However, if the genetically modified products offer benefits that traditional products can not offer, perceived benefits outweigh the perceived risks related to genetic engineering technology.

When consumers have little motivation to process biotechnology information, or little ability to understand it, or no time to digest such information, their opinions will be influenced by the peripheral route to persuasion. Sometimes this happens simply because the information is either too detailed, or because it is too general to integrate into a belief system. In other words, it is either too complex or too general. For instance, while risks from microbiological hazards are reported in quantitative terms (i.e., number of occurrences and percentage increase or decrease), risks from food applications are generally stated using unqualified terms such as "bad" and "thus should be avoided".

This lack of access to information combined with lack of ability to process complex biotechnology information is likely to lead most consumers to engage in more heuristic peripheral processing when forming attitudes toward biotechnology. When the peripheral route is activated, the focus is not on message arguments but non-message cues, such as public opinion, sound bites, emotions generated by advertising, labeling, or the credibility of spokespeople or endorsers.

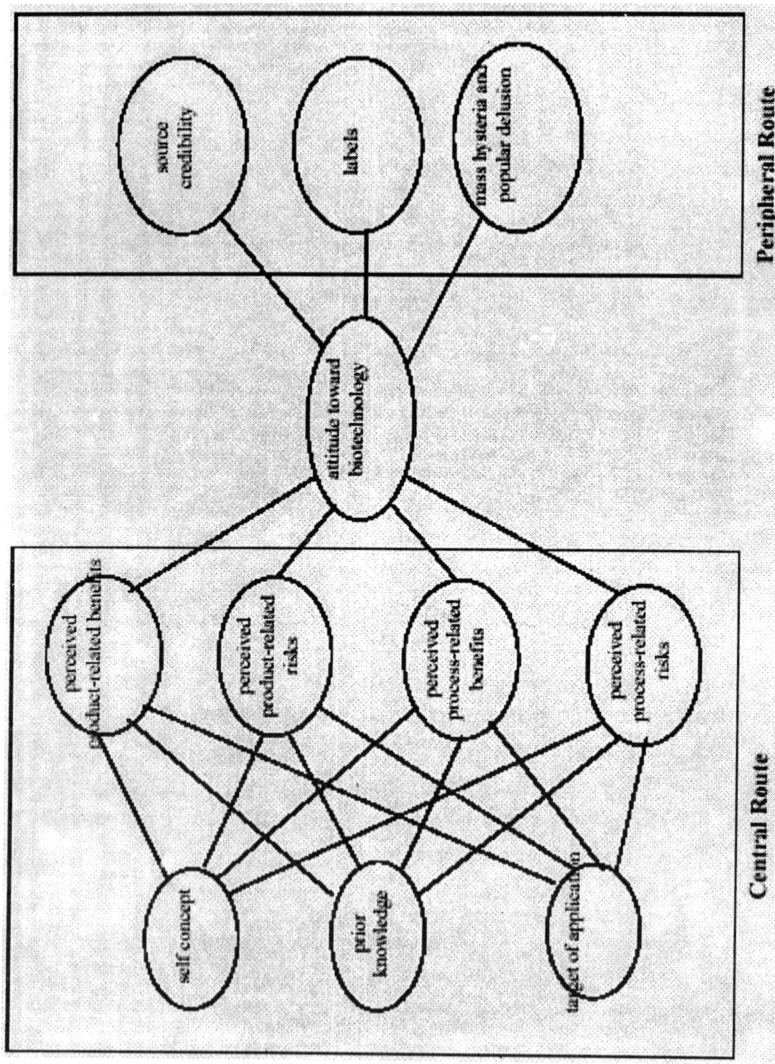

Figure 1. Determinants of Consumer Attitude Toward Biotechnology

This general notion that people can be aware of an issue without having specific knowledge of it is well-supported. Sheehy, Legault, and Ireland reported that consumers, even highly educated ones, had little or no knowledge of biotechnology. However, their awareness of biotechnology, defined as "having heard of the term," was substantially high. This high awareness/low knowledge scenario is particularly common with biotechnology (particularly genetic engineering) because it is new, complex, and not evident to the casual observer. Therefore, consumers can be aware of a biotechnology application while making no associations between it and the genetic engineering process that created the novel characteristics.

Marketing Fallacies

Regardless of one's position, effectively educating consumers about the benefits and risks of biotechnology, necessitates having the proper assumptions about how consumers learn. To date, both opponents and proponents of biotechnology are operating under assumptions about consumers that limit the effectiveness in communicating their message to the public. To be more effective, each of these assumptions must be revised.

Proponent Fallacy 1. The Biotechnology Issue will "Blow Over"

Should the earlier metaphor of biotechnology being the Model T of the 21st century prove right, it will only be because it stays on the road long enough to outlast resistance. Some highly visible companies have bowed to biotech pressures to source select non-GMO food.

A content analysis of stories about biotechnology have shown a balanced reporting of the topic over the past 10 years. Yet, the earlier years tended to be more positive than the later years. The common hope that this is a topic that will "blow over" was an fatal error made by the

British. In 1994, public sentiment toward biotechnology was neutral if not moderately positive.

The industry, therefore, took no real efforts to build public support or enthusiasm for biotechnology because attitudes toward it were improving each and every day. Attitudes were favorable, but neither fully formed nor stable. As a result, when "mad cow disease" became an issue, the industry had not generated the appropriate level of education and support needed to keep the issue in perspective and to keep biotechnology moving forward. This is true with the United States. Perhaps sentiments of the non-vocal majority are moving ahead at a pace that would indicate this whole issue will blow over. The concern, however, is that we are only one "mad cow disease" episode away from losing the biotechnological ground that's been gained.

Because of the highly sensitive nature of this issue, it is critical to "shore up" an understanding of biotechnology and attitudes toward it so that an uncontrollable, or unrelated event would not unnecessarily cause a fatal overreaction. Current negative public opinion about the use of biotechnology, especially in food industry, is largely due to the uncertainty that stems from lack of information by the public. In addition, biotechnology rapid advancement and some of the current safety issues may become non-issues.

In the case that current negative public reaction to the use of biotechnology is temporary, as was the cases for other innovative technologies such as electricity and computers, marketing strategies should focus on counteracting public misperceptions and educating consumers about the benefits of biotechnology. The main task is to accelerate consumer acceptance through the use of more effective persuasion strategies.

Conversely, controversy over safety and ethical issues involved in the use of biotechnology may be a persistent problem and continue to haunt all parties involved even after the majority of the consumers have adopted the products. Adaptation in living organisms and their change is neither completely predictable nor completely controllable. Therefore, in the long run, the biotechnology industry and researchers as well as the government should try to safeguard potential hazards.

First, objective measures of potential risks of hazards involved in biotechnology and biotechnology products must be developed. Without such measures, it will be impossible to convince consumers of the safety of biotechnology. Secondly, legal and selfregulatory protection devices must be strictly reinforced by government and industry. Lastly, and the most importantly, codes of ethics that guards against the potential misuse of biotechnology must be established. Although some groups have issued such codes, candid acknowledgement of the potential hazards of the technology and sincere concerns for human well-being by researchers and the marketers of biotechnology and its products are required to safeguard potential hazards.

Proponent Fallacy 2. Science Sells and Fear Fails:

Consider the case when a person's attitudes have been formed through the peripheral route to persuasion. With relatively low awareness and knowledge of biotechnology along with no established measures of benefits and risks involved in biotechnology applications, his or her biotechnological attitudes will be susceptible to the influence of other people's opinions and reactions and peripheral events. To this person, careful scientific reports and expertly articulated third party testimonials will have little if any direct impact. Indeed, even a judicious FDA endorsement might have little impact compared to a sound bite or to the negative portrayal of genetic engineering applications in a movie.

Perhaps the biggest indicator of how low involvement sells is in looking at how social and ethical taboos play a great role in generating public concerns about biotechnology applications. Widespread negativism related to biotechnology has also been fostered by objections raised by some religious groups and environmentalist organisations. Animal rights activists protest biotechnology on the ground that many genetically modified animals suffer vulnerability to specific diseases as the result of such modifications.

Religious groups oppose the use of biotechnology on the ground that experimenting with lives is like "playing God." These religious and ethical concerns are growing rapidly with the possibility that further advances in gene technology may lead to tampering of human genes in such ways as cloning and gene selection. These groups oppose the use of biotechnology on the bases of social, ethical and religious grounds and provide the public with bases to form an opinion without sufficient scientific knowledge to evaluate the benefits and risks objectively.

The fallacy is that if consumers are given the facts, they will come to the right conclusions. This fallacies takes a "consumers as Computational Machines" approach to understanding consumers. When forming attitudes and judgements about issues, such as biotechnology, basic human tendencies keep consumers from behaving exactly the way the computer programme says they should operate. Some people weigh certain information as more important than other types of information. That is, a well-to-do vegetarian might believes cost savings are less important than how an animal is cared for. A second person concerned with world hunger might focus more on how biotechnology increases the world food supply and slows land commercialisation. A third person might focus on comparing organic gardens of yesteryear to the unknown issues of tomorrow. What is happening is that consumers form their attitudes in different ways of how one combines the information they are objectively given.

Consumers differ significantly in terms of their prior knowledge about biotechnology, current biases toward biotechnology, and personality characteristics. These differences will lead to differences in information processing styles for biotechnology information. It is, therefore, important to develop customised strategies dissemination of biotechnology information tailored to the target audiences. Classification of consumers into several categories based upon multiple variables, such as prior knowledge, information processing style, and current biases toward biotechnology, would provide a sound basis for developing more effective strategies.

Proponent Fallacy 3. Biotechnology is an Industry Association Issue

Biotechnology is not an industry association issue. Biotechnology is a branding issue. It is providing a clear, systematic, vivid, focused message that is potentially important to consumers. In the branding biotechnology war, it is clear who has the upper hand. The powerful visuals that are associated with names such as "FrankenFoods" and "Super Weeds" leave little wonder why the public is able to latch on to such bumper-sticker philosophies of skeptics and be moved.

It is unlikely that trade associations and scientific organisations, and the government can effectively brand biotechnology in a way that leaves it standing clear in a consumer's mind. History indicates the majority of efforts by trade associations, scientific organisations and the government in this regard have not been as effective as hoped for or as claimed. Even the most notable examples (recall the "Got Milk" campaign), won awards, but reportedly contributed little to sales increases among nonusers. If a company is to compete with the "spin" that opponents of biotechnology create, they need to realise that branding biotechnology is something that deserves some of their best marketing minds. It is not something to be outsourced and trusted to a more risk-aversive counterpart.

In terms of trustworthiness, consumer organisations, environmental groups, and biology researchers are generally seen as the most reliable information sources, while the biotechnology industry was the least trusted source. Mistra et al. asked respondents to express the extent of their confidence in comments on food safety issues by different interest groups. The most trusted group consisted of university scientists, followed by independent laboratories and consumer groups. When looking at how to provide communication, it must be realised that the important issue is likely to the least effective less unless supported by consumer groups.

Proponent Fallacy 4. Good for Medicine Means Good for Food

Just because consumers accept biotechnology for medicines, this does not mean they will as readily do it for foods. Generally, consumers are more accepting of genetic engineering and biotechnology related to medicine than that related to food production. These differences may indicate that consumers weigh the perceived benefits and risks differently for genetic engineering depending upon the target of application. Prospect theory provides a possible theoretical framework to understand the differential perception and differential weighting of benefits and risks across application domains. This can be explained by the way the situation is framed - or perceived - by consumers.

People show a risk-taking tendency when the outcome is framed as reduction of a loss, while becoming risk-aversive when the outcome is framed as a gain. In general, biotechnology applications in the medical domain are framed as losses. For example, the benefit of a new medicine developed with biotechnology is generally framed as the improved health condition of an already ill patient. Thus, it is framed as a reduction of losses. The benefit of a food product produced with biotechnology, however, is framed as improved nutrition and quality for a product that already has satisfactory quality and nutrition from a consumer's point of view, and, thus is framed as an increase in gains.

Therefore, consumers are willing to take risks that will reduce a loss, but one loss willing to take risk for something that is classified as a benefit. If the differences in acceptance of biotechnology across application domains are due to differences in how the benefits and risks are framed as proposed by prospect theory, consumer acceptance of food related biotechnology may be improved by framing the benefits in terms of reduction of potential dietary hazards instead of framing them in terms of enhanced nutrition or quality.

Earlier, it was pointed out that confusion regarding criterion for weighing or evaluating benefits and risks involved in biotechnology may lead to deferral of decisions, or the total rejection of biotechnology. Therefore, it is critical to provide consumers with clear information for evaluating biotechnology and biotechnology products. The following is a listing of important consumer education points.

Improving human health conditions through development of better medicines and better food production must take the top priority. The destructive effects of chemical pesticides and wastes are far more devastating to the ecological balance than those of genetically modified living organisms. Ethical concerns regarding the possible misuses of biotechnology should not interfere with the potentially beneficial use of biotechnology as a whole. Biotechnology products on the market have passed safety requirements and can be trusted.

Consumer Analysis and Biotechnology Balance

On the consumer behavior related to biotechnology provides several important insights into understanding the consumer acceptance process of biotechnology and its applications. The differences in prior knowledge about biotechnology and information processing styles among consumers suggest that communication strategies to disseminate biotechnology information and educate consumers should take a stepwise approach. In general, acceptance of biotechnology is a stepwise process. First consumers need to accept the processes of biotechnology, and only after that can they accept specific products themselves. Therefore, having information about and confidence in biotechnology is a necessary condition for acceptance of biotechnology products.

To address this, a hierarchical model of communication strategy is proposed. It will be necessary to disseminate general information such as what biotechnology is and what would be affected by it as the first step of the process. Consumers must have some basic level of knowledge about biotechnology in order to process more specific and detailed information. Next, information about biotechnology used by the specific industries can be understood.

Once the technology or process itself is understood and accepted by consumers, then information about benefits and risks involved with the products can be more effectively conveyed. This way, consumers will be able to develop knowledge basis on which they can make educated decisions regarding specific biotechnology products. Figure 2 illustrates how the hierarchy of communication objectives can be structured in relation to the level of consumers' biotechnology knowledge.

Clearly, consumers have predisposed attitudes towards particular terminology such as "genetic engineering." Regardless of the reasons behind these attitudes, it is important to accommodate the unease invoked by the terminology. Therefore, manufacturers should seek to avoid the use of potentially negative terms either through omission or the use of alternate terminology.

It has also been noted that consumers generally view product labeling as an important source of information when developing attitudes toward biotechnology products. Therefore, labeling and product packaging should reflect the positive aspects of the industry and methodology involved in production. When possible, consumer advocacy organisations and research institutions should be utilised as endorsers for the products or technology as they are viewed as most trustworthy. In the meantime, it will be important to provide consumers a sense of control over their choices.

The use of labeling to provide biotechnology information regarding the production of the products can be important. Labeling of biotechnology products will not only serve an informational function but also serve as a safety signal and reduces concerns about being able to choose wanted products. Biotechnology communication strategies should provide consumers with criteria for evaluating biotechnology products. Consumers will be more comfortable and confident in accepting biotechnology

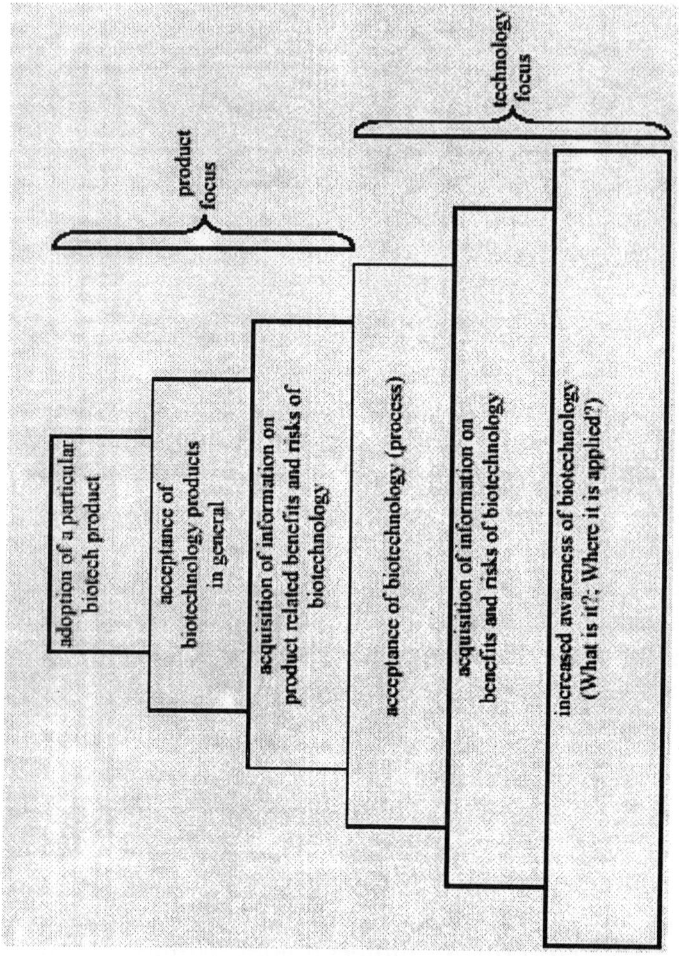

Figure 2. Layered Hierarchy of "Take-aways"

as their confusion about what and how to choose diminishes.

Biotechnology is advancing into the future and some of the current safety issues may become non-issues. However, current public concerns are grounded on what has happened with past misuses of biotechnology. Concerns may partly be due to the fact that living organisms are adaptive and their change is not completely predictable and/or controllable.

Therefore, in the long run, the biotechnology industry and researchers as well as the government should try to safeguard potential hazards. First, objective measures of potential risks of hazards involved in biotechnology and biotechnology products must be developed. Without such measures, it will be impossible to convince consumers of biotechnology safety. Secondly, some legal and self-regulatory protection devices must be put in place by the government and industry. Lastly, and the most importantly, a code of ethics that guards against the potential misuse of biotechnology must be established and adopted by those who participate in the field of biotechnology.

Generally consumers perceive biotechnology information provided by the industry to be the least credible and are most distrustful of an industry regulated safety system. The biotechnology industry is the major provider of biotechnology products that consumers make choices about. Therefore, it is critical for the industry to earn consumers' trust. A self-regulatory effort by the industry may help gain consumer confidence. The biotechnology industry should strive to develop objective measures for risks and benefits of products and establish self-regulated safety measures of the processes used.

Disclosing information on the product labels is also important since it provides consumers control over their choices. Consumers are ill equipped with knowledge, they still desire control in choosing what they eat. The use of product labeling to provide information regarding the biotechnology used to produce the products can be important. Labeling of biotechnology products will not only serve an informational function but also serve as a safety signal and reduces concerns about choice.

The Role of Government as the Safeguard. Despite some doubt regarding the efficiency of government, many focus groups and surveys indicate that consumers believe the government should play an important role in providing regulation and safety protection with respect to biotechnology. These provisions and assurances of safety by the government will contribute to eliminating some of the concerns consumers hold about biotechnology. The government should take the responsibility of setting the direction and pace of development to prevent questionable or premature application of biotechnology. University and Research Institutions.

Universities and other research institutions account for the majority of genetic engineering and biotechnology research and development. Therefore, they are well positioned to play a safety-assurance role as well as provide up to date information on biotechnology advances and applications. While industry sponsorship raises some concerns, the public views these institutions as the most credible and trustworthy source of biotechnology information. Therefore, a more active effort to establish and maintain integrity and impartiality of research by these institutions is important.

When contemplating product positioning in the market as a whole and at the individual store level, marketers should seek to align products with their non-biotech counterparts. This avoids the assignment of a stigma upon the products as being "fake" or "synthetic."

However, this is not as important when targeting biotechnology-savvy markets wherein differentiation techniques may even work as an advantage.

Additionally, efforts should be made to "tie-in" the products with brands and images that are highly regarded and give the impression of being natural. Through the use of brand equity leveraging, innovative promotion, product pairing, etc., the products can achieve an air of familiarity, quality, and conventionality.

Consumers acquire biotechnology information from various sources such as government publications, consumer organisations, research institutes, and the media. Because consumers perceive that there are conflicts among the various information sources and this perception can cause unnecessary confusion leading to rejection or deferral of acceptance of biotechnology products. An integrative and coordinated communication effort by the multiple information sources is essential in increasing consumer acceptance.

Consumers believe that government should play an important role in providing regulation and safety protection with respect to biotechnology. These provisions and assurances of safety by the government will contribute to eliminating some of the negative delusions consumers hold about biotechnology. Universities and other research institutions account for the majority of genetic engineering and biotechnology research and development. Therefore, they are well positioned to play a safety assurance role as well as provide up to date information on biotechnology advances and applications.

While industry sponsorship raises some concerns, the public believes in impartiality of these institutions. Universities and research institutions are major providers of biotechnology advances. The biotechnology industry is the major provider of biotechnology products that consumers make choices. It is critical for the biotechnology industry to earn consumers' trust by implementing self-regulatory efforts. Finally, media has long been established as one of the main information sources for consumers, but less as a source of biotechnology information. The effect of mass media in disseminating biotechnology information has been inhibiting (i.e., distributing news about bio-hazards) rather than facilitating.

6

Genetic Engineering and Food Security

Farmers, young and old, educated and uneducated, have easily taken to the new agronomy. It has been heart-warming to see young college graduates, retired officials, ex-armymen, illiterate peasants and small farmers queuing up to get the new seeds. At least in the Punjab, the divorce between intellect and labour, which has been the bane of our agriculture, is vanishing. While we can and should rejoice about the past achievements of our farmers, scientists, extension workers, and policymakers, there is no room for complacency. We continue to face several problems:

— First, increasing population leads to increased demand for food and reduced per capita availability of arable land and irrigation water.

— Second, improved purchasing power and increased urbanisation lead to higher per capita food grain requirements due to an increased consumption of animal products.

— Third, marine fish production is becoming stagnant.

— Fourth, there is increasing damage to the ecological foundations of agriculture, such as land, water, forests, biodiversity, and the atmosphere, and there are distinct possibilities for adverse changes in climate and sea level.

— Finally, while dramatic new technological developments are taking place, particularly in biotechnology, environmental and social implications are yet to be fully understood.

Because land and water for agriculture are diminishing resources, there is no option but to produce more food and other agricultural commodities from less arable land and irrigation water. In other words, the need for more food has to be met through higher yields per units of land, water, energy and time. Scientific progress on the farms, as an "ever-green revolution," must emphasise that the productivity advance is sustainable over time since it is rooted in the principles of ecology, economics, social and gender equity, and employment generation.

The dimensions of the challenges faced by those involved in developing scientific strategies and public policies for sustainable food security are best defined in some statistics on India, which now has a population of one billion. In global terms, India today has 16 percent of human population, 15 percent of farm animal population, 2 percent of the geographical area, 1 percent of rainfall, 0.5 percent of forests, and 0.5 percent of grazing land. The Green Revolution has so far helped to keep the rate of growth in food production above the population growth rate. The Green Revolution was, however, the result of public good research, supported by public funds. The emerging gene revolution, by contrast, is spearheaded by proprietary science and can come under monopolistic control.

Gene Revolution

Mendel's laws of genetics were rediscovered in 1900.

Mendel had published his work on inheritance patterns in pea in 1865, but it took 35 years for others to grasp their significance. Since 1900, we have witnessed steady progress in our understanding of the genetic makeup of all living organisms ranging from microbes to man. A major step in human control over genetic traits was taken in the 1920s when Muller and Stadler discovered that radiation can induce mutations in animals and plants.

In the 1930s and 1940s, several new methods of chromosome and gene manipulation were discovered, such as the use of colchicine to achieve a doubling in chromosome number, commercial exploitation of hybrid vigor in maize and other crops, use of chemicals such as nitrogen mustard and ethyl methane sulphonate to induce mutations and techniques like tissue culture and embryo rescue to get viable hybrids from distantly related species. The double helix structure of DNA (deoxyribonucleic acid), the chemical substance of heredity, was discovered in 1953 by James Watson and Francis Crick. This triggered explosive progress in every field of genetics.

At the end of the 20th century, we see a rapid transition from Mendelian to molecular genetic applications in agriculture, medicine, and industry. This brief capsule of genetic progress from 1900 to 1999 adequately stresses that knowledge and discovery represent a continuum, with each generation taking our understanding of the complex web of life to a higher level. It would therefore be wrong to worship or discard experimental tools or scientific innovations because they are either old or new. Just as it took 35 years for biologists to understand the significance of Mendel's work, it may take a couple of decades more to understand fully the benefits and risks associated with genetically improved foods. It would be prudent to apply scientific and precautionary principles in areas of human health and environmental safety.

The 1990s have seen dramatic advances in our understanding of how biological organisms function at the molecular level, as well as in our ability to analyze, understand, and manipulate DNA molecules, the biological material from which the genes in all organisms are made. The entire process has been accelerated by the Human Genome Project, which has poured substantial resources into the development of new technologies to work with human genes. The same technologies are directly applicable to all other organisms, including plants. Thus, the new scientific discipline of genomics has arisen, which has contributed to powerful new approaches in agriculture and medicine, and has helped to promote the biotechnology industry.

Several large corporations in Europe and the United States have made major investments to adapt these technologies to produce new plant varieties of agricultural importance for large-scale commercial agriculture. The same technologies have equally important potential applications to address food security and poverty of people in developing countries. Work in India has shown that genetic modification can do immense good in agriculture and food security.

The 21st century may witness changes in temperature, precipitation, sea level, and ultraviolet b radiation, as a result of global warming. These changes in climate are expected to adversely affect India and sub-Saharan Africa. All human-induced calamities affect adversely the poor nations and the poorest among all nations the most. This led us to initiate an anticipatory research programme to breed salt-tolerant varieties of rice and other crop plants in coastal areas, in order to prepare for seawater intrusion into farmland as a result of an eventual rise in sea level. The donor of salt tolerance was a mangrove species belonging to the family Rhizophoraceae. Transferring genes for tolerance to salinity from mangrove tree species to rice or tobacco is an impossible task without recourse to recombinant DNA experiments. This demonstrates the immense benefits that can accrue from genomics and molecular breeding.

In industrial countries, the major concerns relate to the impact of genetically improved organisms (GIOs) on human health and the environment. These food and environmental safety concerns have been well documented and are widely known. The food and environmental scientists of developing countries are equally concerned about the food and environmental safety aspects of GIOs.

There are other issues of concern to the general public in India. First, India is a land of small farm holdings. There are now 106 million operational holdings in the country, and about 75 percent of them are one hectare or less. India has 25 percent of the global farming community, and farming provides a livelihood to nearly 66 percent of the population. There is concern that expansion of proprietary science and shrinking of "public good" research supported from public funds may lead to a situation where the technologies of the future remain in the hands of a few transnational corporations.

Only resource-rich farmers may have access to them, thereby widening further the gap between the rich and poor. This could accelerate social disintegration. Second, monopolistic control over crop varieties could lead to a situation where large areas are covered by very few genetic strains or hybrids. It is well known that genetic homogeneity enhances genetic vulnerability to biotic and abiotic stresses. Biotechnology companies are therefore recommending resistance management strategies, such as growing 30-40 percent non-Bt (Bacillus thuringiensis) corn with Bt-corn.

What will happen to the livelihood of farm men and women operating smallholdings with institutional credit and with no crop insurance, if GM corn, soybean, rice, potato or other crops are affected by serious diseases as a result of the breakdown of resistance? Will the companies agree to compensate them for the loss? This problem could become even more serious if companies incorporate genetic use restriction mechanisms, known popularly as "terminator" genes in the new varieties. Small farmers could then experience "genetic enslavement" since their agricultural destiny could be in the hands of a few companies if they have to purchase new seeds each year, similar to conventional hybrid seed.

A third issue relates to the potential impact of GM foods on biodiversity. This has two dimensions. The first deals with the replacement of numerous local cultivars with one or two new varieties, which could lead to genetic erosion. Modernisation of agriculture has resulted in a narrowing of the base of food security, both in terms of the number of species constituting the food basket and the number of genetic strains cultivated. Local cultivars have often been the donors of many useful traits, including resistance to pests and diseases. Under small farm conditions, every farm is a genetic garden, comprising several annual and perennial crops, and several varieties of each crop. The need of the hour is to enlarge the food basket and not shrink it further.

The second dimension is equity in benefit sharing between biotechnologists and the primary conservers of genetic resources and the holders of traditional knowledge. The primary conservers have so far remained poor, while those who use their knowledge (for example, the medicinal properties of plants) and material become rich. This has resulted in accusations of biopiracy. It is time that genetic engineers and others promote and find ways to implement genuine biopartnerships with the holders of indig-enous knowledge and traditional conservers of genetic variability, based on principles of ethics and equity in benefit sharing.

Principles of bioethics, biosafety, biodiversity conservation, and biopartnerships, there will be serious public concern in India, as well as many other developing countries, about the ultimate nutritional, social, ecological, and economic consequences of replacing numerous local varieties with a few new genetically improved crop varieties. To derive benefits from genetic engineering without undue risks, every nation should set up a multistakeholder Commission for Genetic Modification.

Ecotechnology Revolution

Knowledge is a continuum. There is much to learn from the past in terms of the ecological and social sustainability of technologies. At the same time, new developments have opened up new opportunities to develop technologies that can lead to high productivity without adverse impact on the natural resources base. Blending traditional and frontier technologies leads to the birth of ecotechnologies with combined strength in economics, ecology, equity, employment, and energy.

In water harvesting and sustainable use, for example, there are many lessons to be learned from the past. In the desert area of Rajasthan, India, drinking water is available even in areas with 100 mm annual rainfall, largely because women are continuing to harvest water in simple structures called kunds. In contrast, drinking water is scarce during summer months in some parts of northeast India, with an annual rainfall of 15,000 mm. There is need therefore to conserve traditional wisdom and practices, which are tending to become extinct.

The decision of the World Intellectual Property Organisation (WIPO) to explore the intellectual property needs, rights, and expectations of holders of traditional knowledge, innovations, and culture is an important step in widening the concept of intellectual property rights (IPR). Principles of ethics and equity demand that this invaluable component of IPR be included when the TRIPs (Trade-related Intellectual Property Rights) agreement of the World Trade Organisation (WTO) comes up for review.

FAO has been a pioneer in the recognition of the contributions of farm families in genetic resources conservation and enhancement by promoting the concept of Farmers' Rights. Like WIPO, UPOV (Union for the Protection of New Varieties of Crops) should also undertake the task of preparing an integrated concept of breeders' and farmers' rights and assisting countries in developing equitable and effective sui generis systems for the protection of new plant varieties, as is required for all members of WTO.

The 20th century produced an impressive array of accomplishments in nearly every field of science and technology. The last part of the century was particularly rich in innovations in biotechnology, and information and space technologies. Such advances have had a beneficial impact on human food and health security. The global population was only 940 million in 1798 when Malthus expressed his apprehensions about human capacity to achieve a balance between food production and population. Human numbers reached 6 billion in 1999, and once in every12 years another billion will be added, if current growth rates continue in developing countries.

Science-based technologies supported by appropriate public policies are responsible for food famines becoming rare. The famine of food at the level of an individual today is mostly due to inadequate purchasing power arising from a famine of jobs or employment opportunities. In spite of an impressive stockpile of scientific discoveries and technological innovations, poverty and social and gender inequities are increasing.

According to the World Bank, 1.3 billion people lived on less than US$1 per day and another 3 billion lived on less than US$2 per day in 1993. Nearly 1.5 billion of the world population of 6 billion will live in severe poverty at the dawn of the new millennium. Illiteracy, particularly among women, is still high in many developing countries. It is not only in opportunities for education that children of many developing countries remain handicapped, but even more alarming, in opportunities for the full expression of their innate genetic potential for physical and mental development.

Between 25 and 50 percent of children born in South Asian countries are characterised by low birth weight (LBW), caused by maternal and fetal undernutrition and malnutrition. The UN Commission on Nutrition in a recent report has warned about the serious consequences of LBW for both brain development in the child, as well as the level of health in later life. New technologies supported by appropriate services and public policies have helped to prove doomsday predictions wrong, and have led to the agricultural revolution (the Green Revolution) becoming one of the most significant of the scientific and socially meaningful events of the 20th century.

Four thousand years of wheat cultivation led to Indian farmers producing 6 million metric tons of wheat in 1947. The Green Revolution in wheat helped to surpass in 4 years the production accomplishments of the preceding 4000 years, thus illustrating the power of synergy between science and public policy. There are uncommon opportunities now to harness the power of such synergy to address contemporary development issues such as the growing rich-poor divide, feminisation of poverty, famine of jobs, human numbers exceeding the population-supporting capacity of ecosystems, climate change, and loss of forests and biodiversity. Whether in economics or in ecology, experience has shown that a trickle-down approach does not work. Fortunately, modern information technology provides opportunities to reach the unreached.

Virtual colleges, computeraided and internet-connected, linking scientists and women and men living in poverty can be established at local, national, and global levels to launch a knowledge and skill revolution. This will help to create better awareness of the benefits and risks associated with genetically improved organisms, so that both farmers and consumers will get better insights into the processes leading to the creation of novel genetic combinations.

The future of small farm families will depend on precision agriculture, which involves the use of the right inputs at the right time and in the right way. Biotechnology will play an important role in the major components of precision farming: integrated gene management, soil health care, efficient water management, integrated pest management, management, integrated nutrient supply, and efficient postharvest management. Ecotechnology-based precision farming can help to cut costs, enhance marketable surplus, and eliminate ecological risks.

7

Food Safety and the Environment

Though the practical applications of environmental biotechnology have generally lagged behind those in medical and plant biotechnology, many scientists believe that biotechnology can play an important part in developing sustainable solutions to the problems of air, soil, and water pollution, and waste treatment and reduction. On the opposite side, concerns have been raised about the potential impact of genetically modified organisms (GMOs) on the environment, particularly biological diversity.

In considering the risks associated with genetic modification, it is important to look also at the risks posed by conventional farming (including organic farming) and food production methods. GM tomatoes, potatoes, and other fruit and vegetable products are on the market in the United States. Tomato pure´e made from genetically modified tomatoes was on sale in two British supermarket chains from 1996 to 1999. A wide range of GM crops have been approved for cultivation in the United States.

In the European Union, maize modified for herbicide tolerance and insect resistance has received approval for animal feed and for processing into starch. Soybeans modified for herbicide tolerance have also been approved for food and animal feed uses. Soybean products such as oil, flour, and lecithin are used in a wide variety of processed foods. In many cases, however, the food products which include the derivatives of such GM crops among their ingredients are processed in ways which remove all or most of the modified DNA or protein. Even where modified DNA or protein survive in the final food product, the amount of genetically modified material present is generally small, denatured or otherwise destroyed in biological function. Gene-sized pieces of DNA decay during storage and are readily destroyed in food processing and cooking.

There is broad agreement among most scientists, among food regulatory agencies in this country and other EU and OECD member states, and among international bodies such as the World Health Organisation and the UN Food and Agricultural Organisation that the GM foods currently on the market are as safe as their conventional counterparts. These bodies are independent of the biotechnology industry, have a primary role of safeguarding public health, and possess high levels of scientific expertise among their staff. It is a measure of how much public confidence in Europe has been rocked by the BSE crisis that repeated assurances from independent scientists and regulatory bodies that GM foods are safe to eat appear to have had relatively little effect in easing public concerns.

The prevailing scientific view that current GM food products present no threat to human health rests on several grounds. The cases of genetic modification of crops and foods to date have typically involved the transfer of two or three genes with specific, well-understood functions. This is a more precise and controlled process than the genetic transfers undertaken in the course of conventional breeding which typically involve the random switching of tens of thousands of genes. This includes an assessment of the implications for human health which focuses on the gene product and its function, and addresses both the intentional and unintended effects that may result from the genetic modification of the food source.

Toxicity

Toxins are poisonous substances produced by living organisms; plants, for example, commonly produce them as protection against insects and other pests. Though in most cases such toxins are produced at a low enough level to enable humans and animals to tolerate them, some widely used foods present hazards of varying degrees of seriousness. Several varieties of bean are unsafe to eat raw. Some types of mushroom are poisonous, as are rhubarb leaves. Though the toxic genes in potatoes are generally not expressed, it is still not advisable to eat them raw in any quantity.All parts of the tomato plant except the actual fruit are toxic.

In the past, conventional breeding methods have occasionally transferred deleterious characteristics from one variety to another in a way which has resulted in a threat to human health. In the mid-1980s, for example, celery growers in the United States introduced a new strain of insect-resistant celery which promised substantially higher yields to growers. After people handling the celery developed severe skin rashes, however, it was discovered that the plant was shedding psoralens, natural chemicals which become irritants when exposed to sunlight.

In the 1960s, a new variety of potato, the American Lenape, proved to have a burning flavour caused by high levels of glycoalkaloid toxins. Incidents such as these have led one scientist to comment that 'many of the nightmares predicted for genetically engineered crops have already happened' as a result of conventional artificial breeding.In the assessment of GM foods, toxicity is one of the key traits investigated by both biotechnology companies and regulatory bodies. In the unlikely event that a GM food plant was developed which expressed toxins harmful to humans, it would not receive regulatory approval.

Allergenicity

Allergens are substances that provoke an allergic reaction on exposure by consumption or contact. They act where they first come in contact with the body - generally either the respiratory system, the gastrointestinal tract, or the skin. Their effects range from the fatal to the relatively mild and are usually specific to individuals. No substance is allergenic to everybody, and substances used by millions may be allergenic to just a few. Allergens can be airborne-substances (pollens, dust etc), infectious agents (bacteria, fungi, parasites), foods (peanuts, shellfish, strawberries, milk products, eggs etc), as well as chemicals, dyes and other substances.

Some foodstuffs contain several proteins which are commonly found to be allergenic; wheat and milk, for example, each contain around 20 different allergenic proteins. In addition to allergies proper, there can be other adverse reactions to food for which standard allergy tests prove negative. These are generally forms of food intolerance, i.e. where the body is unable to metabolise the substance. This kind of intolerance is a relatively common response in some individuals to products such as flour and gluten, food colourings and preservatives, and dairy produce.

Recent decades have seen a sharp rise in all forms of allergy - a rise which began long before the introduction of genetically modified crops and foods. Changes in diet, increased exposure to chemicals, and a sanitised way of life which protects people from germs are believed to be among the factors that have contributed to this increase.

Concerns have been expressed that the introduction of new genes into plants might unintentionally introduce proteins capable of causing allergic reactions. The main safeguard against the incorporation of allergenic traits in GM foods lies in the detailed analysis, assessment, and evaluation to which these products are subject before they can be put on the market. A range of tests are carried out to identify possible allergens before they are inserted into plants.

If the foreign DNA comes from a known allergenic food such as peanuts or shellfish, the GM plant is subject to immunological tests. If it does not come from such a source, its amino acid sequence is studied to check for similarities with the sequences of known allergens. The EU Novel Food Regulation also requires that products be clearly labelled if they contain genes that may result in allergenicity, particularly if such genes would not normally be expected to occur in the food.

The only known case where an allergenic protein capable of causing adverse reactions was introduced into a plant by means of genetic modification was detected by tests at an early stage.In the early 1990s, scientists at an American seed company added a gene taken from brazil nuts to soya beans to enhance their nutritional content. At the company's request, an allergy specialist studied the likely impact of the modified soya on patients allergic to brazil nuts. When it was discovered that it was likely to trigger serious attacks in this group of people, the development of the plant was abandoned.

The scope for using genetic engineering to remove allergens from foods may also grow in the future. Researchers in Japan, for example, have reduced the level of the main allergenic protein in rice by inserting an anti-sense gene to block its production in the plant.

Antibiotic Resistant Marker Genes

The transfer of genes from one organism to another is a complex process, and one whose effectiveness is quite variable. When a new gene is inserted into an organism, therefore, researchers will often also insert a 'marker' gene to assist them in ascertaining whether the transferred genes have been successfully incorporated in the host organism. Genes for resistance to a range to antibiotics have been used as selectable markers since the late 1980s. Plants containing a gene for resistance to an antibiotic such as kanamycin will grow on material which contains that antibiotic, whereas plants which have not incorporated the transferred gene will not do so. This makes it possible to screen out plants which have failed to take up the new genes.

The use of antibiotic-resistant genes in GM plants for human consumption has given rise to fears that these genes could be transferred to bacteria present in the human stomach, thereby making them resistant to commonly prescribed antibiotics. Fears have also been expressed that GM plants incorporating antibiotic resistance genes which are used for animal feed could lead to the spread of antibiotic resistance to humans through the transfer of resistant bacteria to those in contact with farm animals.

Though the possibility of transfers of these kinds cannot be ruled out, it should be noted that, despite extensive research, there has been no recorded case of antibiotic resistance being spread through the transfer of antibiotic resistant marker genes to animals or humans.This kind of transfer would be an example of horizontal gene transfer which, even if it once occurred, is very rare in plants and animals. In 1997, the EU granted marketing approval for a GM maize incorporating a gene for resistance to the antibiotic ampicillin for use as animal feed and in the production of starch.

In the view of the European Commission's scientific advisory committee, resistance to this antibiotic was already widespread on farms. It was considered improbable that the DNA would survive intact in the animals' gut to be picked up by bacteria, while any processing of the maize into starch would degrade the modified DNA so that it was no longer functional.In 1998, an application for marketing approval for a potato modified to produce extra starch which contained an antibiotic resistance gene was withdrawn following objections by member states.

While stating that the risk of horizontal gene transfer is extremely low, the Food Safety Authority of Ireland has recommended that selectable markers based on antibiotic resistant genes should be avoided. Alternative marker systems which do not use antibiotic resistance have been developed, and it is unlikely that antibiotic resistant genes will be used as markers in new GM food products. It is also possible to remove marker genes from GMOs after organisms which have incorporated new genes have been identified.

In recent decades, resistance to antibiotics has grown to what many scientists consider an alarming extent due to factors such as the over-prescribing of antibiotics for minor illnesses and their widespread use in modern farming. At EU level, member states have been advised to clamp down on over-the-counter sales of antibiotics and to exert greater control of antibiotic use in human medicine, veterinary medicine, animal feedstuffs, and food.

In the future, enhanced knowledge of genomics and the increased sophistication of GM technology are likely to permit the rewriting of entire metabolic pathways in order to modify several plant traits in tandem. In the case of GM foods, it has been suggested that a possible increase in the use of genes without a history of food use may give rise to a greater risk of toxicity or allergenicity.These developments will necessitate more sophisticated analytical techniques to test the safety of GM crops and foods, and require that regulatory systems keep fully abreast of developments in food applications of genetic engineering.

GMOs and the Environment

Another main area of concern raised by genetic modification is that of the possible effects on the environment. Sharp differences are apparent between, on the one hand, the biotechnology industry which sees genetic modification as an environmentally friendly technology that will lower the use of harmful pesticides and, on the other, environmental groups which believe that it will lead to 'genetic pollution' and inflict a range of direct and indirect damage on the natural world.

Each of the main applications of genetic modification in agricultural biotechnology to date - the insertion of genes for herbicide-tolerance and insect-resistance respectively in a range of plants - have given rise to concerns about their impact on the environment and biodiversity.

Herbicide-Tolerant Crops

Weeds compete with crops for moisture, nutrients and light and can badly affect crop yields and quality.Though herbicides have been extensively used over the past half-century to combat weeds, the more effective they are in doing so the more likely they are to cause collateral damage to crops. This has made the development of crops tolerant to herbicides an obvious goal for biotechnology firms and, to date, herbicide tolerance has been the characteristic most commonly engineered into transgenic crops. Though tolerance has now been engineered for most of the main herbicide groups, the most common are crops modified for tolerance to products containing glyphosphate - a broad-spectrum, non-selective, post-emergent herbicide which can be employed to control most of the major weed species in crops.

Starting with soybeans, tolerance genes for glyphosphate have been incorporated into a range of other crops-including maize, cotton, sugar beet, oilseed rape, and tobacco. In 1999, for example, it is estimated that 50 per cent or more of total soybean and cotton acreage in the United States was accounted for by varieties genetically engineered for glyphosphate tolerance.

Herbicide-Tolerant Crops and Herbicide Usage

Some commentators have argued that the purpose of herbicide-tolerant crops is to enable farmers to spray as much herbicide as they want whenever they want without having to worry about damaging their crops. They maintain consequently that the development of these crops has simply been a device by agrochemical companies aimed at selling greater quantities of pesticide.In response, the companies contend that the planting of herbicide-tolerant crops will require fewer applications of more effective, broad-spectrum herbicides. In support of this claim, they ask why farmers paying higher seed prices for GM seed, together with a technology fee, would want to incur further costs by using greater quantities of herbicide.

Many agronomists would also contend that there are environmental benefits from an increase in glyphospate use and a reduction in the use of other herbicide treatments. Glyphosphate herbicides bind tightly to soil particles and, over time, break down naturally in the soil into naturally occurring components, such as carbon dioxide. This contrasts with soil-acting herbicides which persist in the soil and are liable to leach into water systems with consequent adverse effects.

Herbicide-tolerant crops are also compatible with 'no till' methods which help preserve top soil. To the extent that less spraying is required, moreover, tractors have to pass over fields fewer times with less compaction and damage to soil. Concern has also been expressed about the indirect effects of herbicideresistant crops on the environment as a result, first, of gene escape and the possible transfer of GM crop traits to weeds and non-GM crops and, second, of their impact on wildlife habitats.

Gene Transfer from GM Plants

Concerns about the escape or transfer of genes from herbicide-tolerant plants take several forms. First, the possibility that herbicide resistance could be spread from GM crops to weeds thereby creating herbicideresistant weeds, or what are sometimes termed 'superweeds'. Second, the possibility that seeds shed during the harvesting of herbicide-tolerant crops could grow as weeds in future crops. Third, the effects of the possible transfer of genes from GM crops to non-GM crops on organic farmers and farmers committed to growing conventional crops. These issues raise two main questions. First, how likely are gene escapes and transfers? Second, how serious are the resultant problems if such escapes and transfers occur?

The Likelihood of Gene Transfers

Pollen from GM crops can be transported varying distances by wind, insects, or other means. Though cross-pollination of GM and non-GM crops and plants is possible, the barriers to its occurrence are greater than is sometimes appreciated. Gene transfer can only occur, first, where a GM crop plant has sexually compatible relatives in the area in which it is grown. GM maize and potatoes, for example, have no wild or weed relatives in Europe, while oilseed rape and sugar beet do. Second, such transfers are highly unlikely among inbreeding crop species such as rice and soya in which crosses occur only between closely related parent plants, as they also are among crops such as cereals which are 99 per cent self-pollinating with only 1 per cent out-crossing.

The main risk of gene transfer from GM to non-GM crops arises where compatible relatives exist close to out-breeding GM crop species such as oilseed rape. It is important to realize that, even where this risk is present, a number of steps must take occur before any transfer of modified genes can take place:In view particularly of 'the relative lack of research data on large-scale releases of GM crops such as oilseed rape under European and Irish conditions, a prudent approach should be taken to the assessment and management of such risks.

Gene Escape and Weeds

There are also significant obstacles acting against plants modified for herbicide resistance or other traits subsequently becoming weeds themselves. Even if gene transfer from GM plants to wild relatives occurs and results in viable offspring, the effects are unlikely to be long-lived in most cases. With limited exceptions which arise to be addressed in the context of specific risk assessment and risk management strategies, crosses between cultivated crop species and wild relatives will generally not prove competitive in the wild.

Outside farm areas, such transfers would not present problems in areas such as woodland, though they could cause difficulties in areas dependent on chemical control such as roadside verges, railway tracks and airport runways.In that event, they, like the GM crops from which they derived, would be resistant only to a restricted range of herbicides and could be attacked with alternative herbicide products.

Gene Transfer to Non-GM Crops

The likelihood of gene transfer to non-GM crops depends, like that to wild plants, on the biology of the crop species and its location. The possibility of such transfer poses risks for organic farmers and farmers who wish to remain GM-free, and may present a threat to their status and viability. These concerns are not dissimilar to those which organic farmers have about the accidental spread of fertilisers, pesticides, or other chemicals onto their land from neighbouring farms.

As the commercial cultivation of GM crops compatible for breeding purposes with crops grown organically draws closer, effective procedures, such as prescribed isolation distances, will be needed to minimise the risk of gene transfers. It should be noted, in conclusion, that the genetic engineering of plant sterility might offer an effective, if controversial, solution to the spread of transgenes through pollen dispersal.

So-called 'terminator technology' being developed by US researchers enables the seeds of GM crops to be prevented from germinating. As the use of this technology would require farmers to buy new seed every year, its implications for poorer farmers in developing countries have been the subject of particular criticism. Though it is certainly open to objection on these grounds, its potential for environmentally benign applications in preventing unwanted gene flow should not be discounted.

GM Material in Seed of Conventional Varieties

The risk of gene transfer from GM to non-GM crops was underlined by the disclosure in May 2000 of the presence of GM material in seed of two conventional oilseed rape varieties imported from Canada and sown commercially in 1999 and 2000 on farms in seven EU countries on an area in excess of 15,000 hectares. It is understood that this came about as a result of cross-pollination in Canada between the conventional oilseed rape and a variety genetically modified for herbicide tolerance.

The affected seed varieties were also included in small scale field trials undertaken by the Department of Agriculture, Food and Rural Development in 1997 and 1998. Such trials are carried out to test new crop varieties for potential value for cultivation and use to Irish agriculture. The affected varieties were submitted and accepted for testing on the basis that they were conventionally bred. The total area sown in each of the two years in which trials took place in this country was less than one-tenth of an acre.

Though the GM presence in the seed batches in question was at a relatively low level (between 1 and 2.4 per cent) and posed no threat to human health, the incident aroused understandable unease in this and other EU countries. The Minister for Agriculture, Food and Rural Development stated that the submission, accidentally or otherwise, of contaminated seed for variety trial evaluation in this country was a source of 'great concern' and indicated that the Irish Government was determined to prevent its recurrence.The Department of Agriculture, Food and Rural Development, in consultation with the Environmental Protection Agency, is to establish an appropriate monitoring and control system to ensure that contaminated seed is not imported in the future.

Herbicide-Tolerant Crops and Biodiversity

Herbicide-tolerant crops may prove to reduce total herbicide use, but their appeal to farmers is based squarely on their greater effectiveness in destroying weeds. It has been claimed as a result that their cultivation will inevitably have an adverse effect on the availability of habitats for insects and other invertebrates. It will also affect the organisms associated with the root systems of weed species. This will in turn lessen the supply of food for the predators that feed on these insects and organisms. Others take a different view, arguing that, as a general rule, undisturbed ecosystems are more diverse and that there is no reason to suggest that a reduction in the biodiversity of hedgerows and other crop surroundings will necessarily result from the cultivation of herbicide-tolerant crops.

In Britain, it is estimated that, in the past twenty years, ten million breeding individuals of ten species of farmland birds have disappeared from the countryside.The intensification of agriculture is generally seen as the cause of these wildlife losses, in particular factors such as the increased use of agrochemical inputs, the introduction of new crop types, increased land drainage, reductions in traditional rotation practices, and hedgerow removals. One U.S. scientist has estimated that pesticides alone may kill up to 67 million birds and between 6 and 14 million fish each year in the United States.Evaluations of the risk to biodiversity from GM crops should consequently take due account of the environmental damage caused by the corresponding conventional crops.

Environmental Effects of Insect-Resistant GM Crops

Plant-eating insects can cause great damage to crops; the European corn borer, for example, can destroy up to 20 per cent of maize yields. Over the past half-century, insecticides have become the main weapon of choice against insect pests. In the 1960s, concerns about their adverse health and environmental effects led to the prohibition or restriction of the most powerful broad spectrum contact poisons such as DDT.

Alternative insecticides have been developed which attack selectively instead of indiscriminately and which break down into non-toxic substances in the environment. Though these are subject to stringent testing, concerns about their effects persist, particularly about residues in food and water and their effects on non-target insects. The development of transgenic crops to enhance resistance to insect pests has been a focus of the agricultural biotechnology sector since the 1980s.

Genes expressing toxins from the soil bacterium Bacillus thuringiensis (Bt) were first engineered into tobacco in the middle of the decade. Bt toxins are highly specific to, and highly effective against, particular groups of insects but are not harmful to other organisms. They are also biodegradable and have been used as commercial insecticides in spray formulations since the 1950s. Though environmentally benign, their utility in spray form has been limited by high production costs and variable operational persistence.

Several varieties of Bt maize, cotton, and potatoes have received regulatory approval in the United States. Bt maize and cotton are estimated to have accounted for 25-30 per cent of the total US acreage of these crops in 1999.A variety of Bt maize which also incorporates a gene for herbicide tolerance has received marketing approval in the EU, but the acreage under cultivation in member states is negligible. Other transgenic crops incorporating Bt toxin genes are currently in development, as are genetically modified plants incorporating other proteins with insecticidal properties such as lectins and protease inhibitors.

Crops genetically modified to incorporate Bt genes, and thus to produce the Bt insecticide inside the plant tissues, offer growers a number of advantages over other insect pest control methods.Unlike insecticide sprays, their utility is not dependent on weather conditions; plant parts difficult to reach by spraying are also protected. Particularly in the case of crops which require heavy applications of insecticide, there are significant potential savings to growers in pesticide, labour and equipment costs.

It has also been claimed that crops modified for insect resistance offer environmental advantages compared with conventional insecticide spraying. Spraying, especially from the air or in windy conditions, tends to be indiscriminate in its impact and often affects non-target plants in the vicinity. Bt toxin produced within the plant by genetic modification only kills insects which eat the plant tissue, and only certain kinds of insects at that because of its specificity. Contamination of ground water around farmland should also be reduced as the Bt toxin is a natural protein which is readily biodegradable into non-toxic substances. Critically, the use of Bt crops should mean that less insecticide is needed as the insecticidal material is contained in the plant tissue.

There are possible risks associated with the modification of crop plants for insect resistance. Concerns have been expressed, first, that insects will evolve resistance to GM plants. This has been a long-standing problem with conventional insecticide treatments. The main counter-measure at present against the threat of enhanced resistance consists of the provision of 'refuges', that is areas of non-GM crops, which can support pest populations that are not resistant to Bt toxins.

The Environmental Protection Agency in the United States insists that, when farmers grow Bt cotton, 4 per cent of the crop must consist of unmodified cotton that has not been treated with any pesticide; if growers wish to spray with other pesticides, 20 per cent of the crop must be non-Bt cotton.In the case of Bt maize, the Agency's new regulations announced in January 2000 provide that unmodified maize must account for at least 20 per cent - and in some cases up to 50 per cent - of the total crop.While such provisions may help to address the problem, some recent research findings have raised questions about the effectiveness of refuges.This underlines the need for continued careful monitoring of the risk of enhanced resistance to Bt crops among target insects.

Effects on Non-Target Insects

The other main concern about insect-resistant crops concerns their possible impact on non-target insects, some of which play an important role in the control of insect pests. On the one hand, the impact of such crops might be expected to prove beneficial because, unlike sprays which may affect all susceptible insects in the field, they should directly affect only insects which attack the plant. On the other hand, GM crops may adversely affect non-target insects by depleting the supply of prey on which they feed. This problem also arises with non-GM pest control measures and gives cause for concern only to the extent that Bt crops are more effective in killing insect pests.

Risks and Benefits

There are risks associated with genetic modification, and these should neither be minimised nor overstated. The risks associated with the current applications of GM technology are capable, in the view, of being adequately assessed and managed by a combination of effective prerelease testing, ongoing research, and post-release monitoring in the context of a modern regulatory framework. The risks presented by future GM products must at all times and in all instances be assessed on a case-by-case basis and in accordance with the step-by-step principle.

The risks associated with modern biotechnology should also be evaluated in the light of the risks posed by conventional farming (including organic farming) and food production methods. Though modern agricultural and food manufacturing methods have provided cheap, safe, and plentiful food in a historically unparalleled way, this has been at the cost of some harm to the natural environment and occasional, limited threats to food safety and human health. Many of the hazards attributed to genetic modification such as antibiotic resistance, allergenicity, and adverse effects on biodiversity were established trends in contemporary societies long before the introduction of GM crops and foods.

The benefits offered by modern biotechnology - and the economic and other costs of missing out on them - must be considered alongside the possible risks. The potential gains they offer individuals and societies across a wide range of areas should not be jeopardised unless there is good cause to believe that they will be outweighed by the harms.

8

Agricultural Biotechnology in Asia

In Asia, several governments are committed to the use of modern biotechnology in agriculture. They have devoted significant human and financial resources to this policy over the past two decades. Some illustrations of current activities in selected countries are given below. The Peoples Republic of China (PRC) accords high priority to biotechnology to increase food production and improve product quality in an environmentally sustainable manner. The PRC has moved quickly to adopt new biotechnologies, particularly genetically modified crops. The country is rapidly increasing its expenditure on biotechnology R&D.

Over 103 genes have been evaluated for improving traits in 47 plant species. New traits have been introduced and evaluated in field tests on rice, wheat, maize, cotton, tomato, pepper, potato, cucumber, papaya, and tobacco. A variety of traits have been targeted. They include resistance to diseases, pests, and herbicides, and quality improvement. Approximately 50 genetically modified varieties have been approved for environmental release, or small-scale field testing in the PRC.

A few new genetically improved varieties have been approved for largescale commercial production. The most widespread are new pest-resistant varieties of cotton that are being widely cultivated by farmers. These were grown commercially by approximately 3 million farmers on approximately 500,000 ha in 2000. Several new products are in the pipeline for potential commercialization.

India has allocated large public resources toward human resources development and infrastructure in biotechnology. In the early 1980s, the Government of India created a Department of Biotechnology to promote the use of new biotechnologies in industry, medicine, and agriculture. Current R&D efforts in India are directed toward increasing agricultural productivity, bioremediation in the environment, medical and industrial biotechnology, and bioinformatics.

R&D priorities in agriculture include new regeneration techniques for the rapid multiplication of citrus, coffee, mangrove, vanilla, and cardamom. Cardamom yield has increased 40 percent through the use of tissue culture. There is substantial private sector participation in biotechnology in India, for example, in the seed sector, the veterinary products sector, and bioinformatics linked with the booming information technology sector. Further, organizations like the M.S. Swaminathan Research Foundation have developed innovative approaches such as the ADB-supported BioVillages programme that is fostering the growth of new income-raising technologies. Emphasis is on developing small-scale bioindustries for women.

Indonesia has placed a high priority on biotechnology over the past 15 years. The Government has designated three National Biotechnology Centers to coordinate R&D in agriculture, medicine, and industrial microbiology. Applications of biotechnology to agriculture are primarily the responsibility of the Agency for Agricultural Research and Development (AARD). A National Committee on

Biotechnology advises the minister in developing guidelines for government policy in the promotion of biotechnology.

In recent years there has been an extensive training programme within Indonesia and abroad to upgrade skills of scientists involved in biotechnological research. In the 1980s, a major World Bank loan of over $100 million financed the creation of three inter-university centers for agricultural, medical, and industrial biotechnology. More recently, a current World Bank loan is financing facilities for agricultural biotechnology within AARD.

Crop improvement efforts using modern biotechnology started in Pakistan in 1985, when a training course was held on recombinant DNA. Work is now concentrated on chickpea, rice, and cotton. Field evaluation is hampered by lack of biosafety regulations. There is some private investment in R&D of agricultural biotechnology. The government controls testing, multiplication, distribution, and biosafety issues for genetically modified crops. Pakistan lacks firm policy and regulations regarding intellectual property rights (IPR) and patents involving biotechnology, and biosafety regulations for GMOs (Zafar 2001).

The Philippines began its modern biotechnology programmes in 1980 with the creation of the National Institutes of Molecular Biology and Biotechnology in Los Baños, with a focus on agricultural biotechnology. In 1997, the Agriculture Fisheries Modernization Act recognized biotechnology as a major strategy to increase agricultural productivity. The Act provided a budget for agricultural biotechnology of almost $20 million annually for the next 7 years.

In 1998, the government funded these five high-level biotechnology research projects to develop:

(i) new varieties of banana resistant to banana bunchy top virus and papaya resistant to ringspot virus,

(ii) delayed ripening papaya and mango,

(iii) insect-resistant maize,

(iv) marker-assisted breeding in coconut, and

(v) coconut oil with high lauric acid content.

Nongovernment organizations (NGOs) and other groups concerned about the safety of GMOs have been vocal in the Philippines. This is affecting field-testing and commercialization of transgenic crops. Products in the regulatory pipeline include new varieties of insect-resistant maize and insect- and disease-resistant rice.

Thailand is focusing on the applications of biotechnology to traditional foods, fruits, and export commodities such as shrimp. R&D priorities are to increase production and reduce production cost on crops such as rice, cassava, sugarcane, rubber, durian, and orchids. An early success in Thailand has been the development of new molecular diagnostics for the diagnosis and control of virus diseases in shrimp. These diseases cost the shrimp export industry over $500 million in lost production in 1996. The development and commercial use of the new diagnostics prevents the loss of an estimated 20-50 percent of annual production, a saving of at least $100 million per year.

There are also active agricultural biotechnology programmes in Bangladesh, Malaysia, Nepal, Sri Lanka, Singapore, and Viet Nam.

Regional and International Programmes

International Agricultural Research Centers

International R&D programmes using modern biotechnology are being conducted by the international agricultural research centers (IARCs), particularly the International Rice Research Institute (IRRI), International Maize and Wheat Improvement Center (CIMMYT), International Crops Research Institute for the Semi-Arid Tropics (ICRISAT), International Livestock Research Institute (ILRI), and International Service for National Agricultural Research (ISNAR). The Center for International Forestry Research also uses biotechnology in the characterization of forest diversity in its Asian programme.

The International Center for Living Aquatic Resources Management is using new technologies in the improvement of fisheries and aquaculture systems. ILRI is initiating a programme on Asian livestock improvement. The CGIAR centers invest approximately $30 million per year in modern biotechnology.

ISNAR Biotechnology Service

The ISNAR Biotechnology Service (IBS), with Japanese support, has been assisting selected Asian countries in developing human resources for managing biotechnology research programmes or institutions. IBS has developed specialized courses to enhance the capacity and competency of managers, focusing on strategy building, priority setting, managing biosafety and regulatory aspects, resource generation and deployment, product delivery, and information sharing as well as the establishment and management of linkages.

International Service for the Acquisition of Agri-biotech Applications

The International Service for the Acquisition of Agri-biotech Applications (ISAAA) is brokering public-private sector partnerships to facilitate technology transfer. It has current regional projects to increase the productivity of maize, papaya, and sweetpotato. ISAAA is also establishing a new Asian knowledge center for crop biotechnology, based in the Philippines, in partnership with CAB International and the Southeast Asian Regional Center for Graduate Study and Research in Agriculture. Its purpose is to make available timely and balanced information on the risks and benefits of crop biotechnology to interested parties in Asia. In doing so, it aims to provide training and study tours highlighting experiences not only in research but also with biosafety and intellectual property issues.

Activities of Funding Agencies

The major external sources of assistance for agricultural biotechnology in the Asia/Pacific region are ADB, Australia, the Rockefeller Foundation, the United Nations, the United States Agency for International Development (USAID), and the World Bank.

Asian Development Bank

ADB has made several strategic and innovative investments in agricultural biotechnology over the past decade. These investments have been in the form of loans and technical assistance projects.

a) Components of Agriculture and Science and Technology Loan Projects

Several governments have requested ADB financial support for human resource development, laboratory facilities, and equipment for agricultural biotechnology programmes. These programmes have been integrating new applications of biotechnology into their conventional agricultural R&D programmes. The applications include the use of new molecular diagnostics for pests and diseases and MAS for crop and livestock breeding. Such components are being supported under ongoing ADB projects in the Philippines, Sri Lanka, and Thailand (Table 3.2). ADB has provided other loans and grants in closely related areas such as in integrated pest management in cotton under the Cotton Development Project in Pakistan.

b. ADB Regional Technical Assistance Projects

ADB has provided regional technical assistance grants for the development of the three regional crop biotechnology networks over the past decade. The networks are the Asian Rice Biotechnology Network (ARBN), initiated in 1993, the Asian Maize Biotechnology Network (AMBIONET), initiated in 1998 and the recently established Asian Semi-Arid Crops Network. These networks have been influential in developing capacity to use new techniques in crop breeding for the major cereal crops (rice, maize) and the crops important in the semi-arid regions (sorghum, pigeonpea, and groundnut). The networks are managed by three IARCs (IRRI, CIMMYT, and ICRISAT) that work with national re-search institutes to implement the programmes on rice, maize, and semi-arid crops.

A recent evaluation of ADB's research investments through the international centers reported favorably on the achievements of the rice and maize biotechnology networks, and recommended further support for this type of research cooperation. The networks provide a platform by which countries and the IARCs can collaborate in the use of the new tools of functional genomics to identify genes that control important traits such as drought and salinity tolerance. This will speed up the breeding of crop varieties with these characteristics that have been difficult to address through conventional breeding. There will be increasing opportunities for cooperation between the networks, as more knowledge is gained on the commonality of genes between species and their control.

Australian Government

Over the last 15 years, ACIAR has entered into more than 100 biotechnology R&D partnerships with at least 10 Asian countries in support of more than 600 active or completed projects. Most of these projects have been implemented through government programmes, but NGOs are becoming increasingly involved. Molecular marker methods for identifying disease- resistant genes and prolificacy in livestock have also been developed. Several projects on biotechnology for rumen manipulation have been carried out.

Crops and forestry work has focused on development of diagnostics for diseases (viral, fungal, mycoplasma, and bacterial) and contaminants in tropical crops, and the application of biofertilizers, bioremediation technology, and biofumigants. Molecular markers have been developed for the improvement of cereals and tree species. In cooperation with IRRI, attempts are being made to develop apomixis systems for rice. Tissue culture methods for micropropagation and conservation of several species, including sweetpotato, taro, tropical fruits, coconut, green tea, and tree species such as mangrove are being developed.

Eight of ACIAR's current or completed projects have included aspects of plant genetic engineering, with target crops being cereals and pulses, groundnut, and several tropical fruits. Target characters include virus resistance and quality defects related to ripening processes. These collaborative projects were developed at the request of the Asian countries, which fully approved regulatory procedures. ACIAR also provides core funding to many CGIAR and other IARCs, a proportion of which is applied to biotechnology R&D.

Australian support has been provided through AusAID for an Association of Southeast Asian Nations

(ASEAN)/Australia regional biotechnology network, mainly concerned with food, microbial, and industrial biotechnology. AusAID also supports the characterization and conservation of genetic resources of taro and forest genetic resources in the South Pacific. AusAID has supported several biotechnology seminars. A major seminar in 1989 at the Australian Academy of Sciences on the potential of agricultural biotechnology in international development reported on the outcomes of a joint study cosponsored by AusAID, ACIAR, ISNAR, and the World Bank.

In the area of human resources, the Crawford Fund has sponsored several master classes in biotechnology for senior policymakers over the past decade.

The Rockefeller Foundation

Operational from 1984-1999, the Rice Biotechnology Network sponsored by the Rockefeller Foundation was very active and successful in the region. The programme's two objectives were to (i) to create biotechnology applications to produce improved rice varieties suited to developing country needs, and (ii) to train scientists in developing countries to use the techniques and adopt them to their own objectives. A network of about 200 senior scientists and 300 trainee scientists from all the major rice producing countries of Asia and a number of industrialized countries participated. The programme was funded at approximately $5 million per year for 15 years.

Another outcome of the programme was the development of promising new technologies for the control of rice pests and diseases and improving the nutritional quality of rice. The work is being continued by national governments, ARBN, and IRRI to do the field evaluation and distribute new rice varieties in Asia.

The Rockefeller Foundation presently concentrates its biotechnology programmes in Asia on drought tolerance in rice and maize. It is also examining innovative means of dealing with access to technologies and IP issues.

United Nations Agencies

The Food and Agriculture Organization (FAO) is giving high priority to biotechnology within its Asia/Pacific regional programmes. During 1989-1993, the United Nations Development Programmeme financed the establishment of biotechnology centers in eight countries (PRC, India, Indonesia, Republic of Korea, Malaysia, Pakistan, Philippines, and Thailand) to share rDNA techniques in animal improvement, embryo transfer, and disease control. In 1994 the United Nations Industrial Development Organization (UNIDO) established an International Center for Genetic Engineering and Biotechnology in New Delhi, India, to assist Asian countries in the applications of biotechnology to important crops of the region.

United States Agency for International Development The USAID is supporting applications of biotechnology through bilateral activities in several countries, including India, Indonesia, and Pakistan. USAID is also providing specific support for biotechnology applications within the programmes of IARCs. This includes support for research by IRP.I on improving the nutritional quality of rice by increasing its vitamin A and iron content, and support for research by the Tata Energy Research Institute on the development of golden mustard.

World Bank

The World Bank has supported the development of infrastructure and human resource development for biotechnology in several Asian countries over the past 15 years. This support has come through loans in the agricultural sector, science and technology, and education. There are currently substantial components for biotechnology within agricultural technology projects in India, Indonesia, and Pakistan.

The Inter University Centers for Biotechnology in Indonesia were built with a $150 million loan in the 1980s. The Republic of Korea also used World Bank loans to develop its infrastructure in biotechnology. The World Bank is also one of the main financial supporters of CGIAR centers. The centers invest about 10 percent of their total annual budget of $340 million in the applications of biotechnology.

Increasing Food Production

Increasing food production has always been the highest agricultural priority in China because of the huge population of the country. It is projected that the population of China will reach 1.6 billion by the year 2030. Demand for food production will increase by at least 60 percent to keep pace with population growth. This rapid population increase and vast urbanization will result in loss of valuable farmland and other natural resources. The only viable approach to increasing food production, therefore, is to increase the productivity of existing farmland.

Statistics show, however, that the total production rates of the major grain crops has been decreasing in the last decade, because yield potentials of the newly released cultivars and hybrids have not been realized. There is also a huge demand for quality improvement of food products, especially the grain quality of cereal crops. Quality improvement of rice, for example, was largely neglected in breeding programmes in recent years. High yield cultivars and hybrids is frequently associated with poor quality; most of the widely used cultivars and hybrids have poor cooking and eating qualities, and thus are disfavored by producers and consumers.

Another major problem is degradation of the environment. We have seen increasingly frequent natural disasters such as floods, drought, insect pests, and diseases, and also expanding areas of soil desertification, salinity, and acidity. Extensive applications of chemicals have created a vicious circle in which the excessive use of the chemicals has resulted in a rapid deterioration of the environment, and this deterioration has made crop production even more dependent on chemicals.

In the last 15 years there have been rapid developments in China in scientific infrastructure and also research programmes in biotechnology and molecular biology of various crop plants. Infrastructure developments include the establishment of National Key Laboratories in the general areas of agricultural biotechnology and crop genetics and breeding, in north, central and south China. These laboratories are well equipped for biotechnology and molecular biology research. In addition, there are open laboratories supported by the Ministry of Agriculture, the Ministry of Education, and the Chinese Academy of Sciences. These laboratories have provided good opportunities for biotechnology research.

Financial Resources

During the same period, regular funding channels were formed at the central government level, which support basic and applied research. This includes the establishment of the National Natural Science Foundation of China and The Chinese Foundation of Agricultural Scientific Research and Education. Major research initiatives and programmes were also established at the state level and by various ministries. The most important programmes for biotechnology R&D are the National Programme on High Technology Development (also known as the 863 Programme) and the National Programme on the Development of Basic Research (also known as 973), both of which included agricultural biotechnology as a major component. Programmes were set up to promote young scientists by awarding special grants from the National Natural Science Foundation, the 863 Programme, and also various ministries. Similar systems, although smaller, were also developed by local governments in many provinces.

International funding channels also opened to Chinese scientists during this period, including those of the Rockefeller Foundation, McKnight Foundation, the International Foundation for Science, and the European Union-China collaboration programmes. The availability of financial support has enhanced research capacity and has promoted the development of young scientists. Some of the programmes have a training component as well.

Scientific Advances

Rapid advances have been made in molecular biology and biotechnology research in China in the 1990s. These include genomic studies in rice and other cereals, development of molecular marker technologies, identification, and mapping and molecular cloning of a large number of agriculturally useful genes. These studies have resulted in powerful tools for crop improvement (for example, marker-assisted selection) that can be applied to develop new cultivars and hybrid parents.

Transformation technologies have also been firmly established in many laboratories for most of the crop species, including major cereal crops such as corn, rice, and wheat that are often considered difficult to transform. Transgenic plants can now be routinely produced for crops such as rice, corn, wheat, cotton, tomato, potato, soybean, rapeseed, and other crops, using Agrobacterium, particle bombardment or other methods. The most up-to-date molecular technologies necessary for varietal development are now in place in China.

Genome mapping and biotechnology research in recent years offer powerful tools in crop improvement including genetic transformation and molecular marker-assisted selection. These techniques have opened enormous opportunities to meet the challenges of food production. These opportunities according to individual traits are described below:

Disease resistance: More than 20 genes for resistance to various plant diseases have been isolated in recent years. Analyses of the DNA sequences indicate that the genes share many structural characteristics in common, despite the fact that diseases are caused by a variety of pathogens such as fungi, bacteria, viruses, and nematodes. The genes were isolated from a wide range of plant species including monocotyledonous and dicotyledonous species including tomato, rice, tobacco, and barley. These have provided a rich source of disease-resistance genes for improving resistance by genetic engineering.

Large numbers of genes have been tagged and mapped using molecular markers in many crop species. Closely linked markers flanking both sides of the genes were identified in many cases. These closely linked markers can be used as the starting points for isolating the genes using the map-based cloning approach. These markers can also be used as selection criteria in breeding programmes to monitor the transfer of the genes, which is referred to as marker-assisted selection. New crop lines with improved resistance have been obtained using both approaches.

Insect resistance: Genes for resistance to various insects have been identified in many crop species and their wild relatives, including gall midge and brown planthopper resistance in rice, and pink borer resistance in cotton. A number of insect resistance genes has also been genetically tagged and mapped using molecular markers. These genes can be directly used in crop breeding programmes using markerassisted selection.

An important strategy in the development of insect resistant crop varieties is utilization of exogenous genes, including genes coding for endotoxin of Bacillus thuringiensis (Bt) and proteinase inhibitors from various sources. Some of the genes have demonstrated strong insecticidal activities under both laboratory and field conditions. Several genes have now been widely used in transformation studies. Many insect-resistant transgenic cotton, corn, and rice plants have been produced from these transformation studies, which have now been advanced to the stage of commercial production.

Large-scale utilization of the insect resistance genes in crop production will not only reduce labor and costs of production, it will also have longterm beneficial effects on the environment. These insect-resistant crops may have a major role to play in sustainable agricultural systems. Tolerance to abiotic stresses: Drought, soil salinity, and acidity are among the most important threats to agricultural production that cause severe yield losses of all major food crops worldwide. In China, the northwest region is prone to drought, so water supply is a major limitation for crop production; in south and central China, soil acidity is a major limiting factor that reduces crop yield; salinity occurs in large areas in the east coastal region.

Drought resistance has been the subject of many studies in several major food crops including rice, corn and sorghum. Although many quantitative trait loci (QTLs), which explain certain genetic variations in drought tolerance in experimental populations, have been identified by molecular marker mapping, they are unlikely to have a major role to play for improving the drought tolerance of crops. There have also been QTL studies on the tolerance of rice to acidic soil conditions, especially with respect to aluminum and ferrous iron toxicity, showing that major gene loci may be involved in increasing the tolerance of rice plants. This may present an opportunity for using genes from rice itself to improve the tolerance of rice varieties to acidic soils. A more promising line of research is the use of gene coding for citrate synthase, the enzyme for biosynthesis of citric acid.

Transgenic sugar beet plants with elevated expression of this gene show an enhanced tolerance to aluminum, and also increased uptake of phosphate in the acidic soil as a result of excretion of citrate. This indicates that genetic engineering may be able to produce plants that can grow better in acidic soil even with reduced application of phosphate fertilizers.

Product quality: Biotechnology may have much to offer in the improvement of product quality. In rice, for example, the poor cooking and eating qualities of high-yielding cultivars and hybrids represent a major problem for rice production in China. Research has established that the cooking and eating qualities are to a large extent dependent on three traits: amylose content, gelatinization temperature, and gel consistency. It was recently shown that all three traits are controlled by the waxy locus located on chromosome 6.

The waxy gene was isolated from maize and rice. Rice plants transformed with the waxy gene both in sense and antisense configurations showed reduced amylose content, thus demonstrating the usefulness of the transgenic approach in improving cooking and eating qualities. Moreover, the waxy locus has also been clearly defined in the molecular linkage map, and markers residing on the waxy locus and closely linked makers that flank the waxy locus on both sides were identified. Thus, improvement of the cooking and eating qualities can therefore be achieved using marker-assisted selection.

Another example is the recent success in engineering the entire biochemical pathway for provitamin A biosynthesis, which significantly enriched vitamin A content in the endosperm of rice grains. This will be a great help to the poor peasant farmers to balance the micronutrients in their diets and hence alleviate malnutrition.

Increasing yield potential: Several of our major crop species have gone through two great leaps in yield increase in the last several decades: increasing harvest index by reducing the height by making use of the semidwarf genes, and utilization of heterosis by producing hybrids. Reduced rates of yield increase have been observed in a number of major food crops in the last 10-15 years.

Increasing yield potential has therefore been a common concern in essentially all crop breeding programmes. Two approaches are follows:

The first approach is called "wild QTLs," in which efforts are devoted to bringing QTLs for yield increase from the wild relatives to enhance the yield of cultivars. The argument for such an approach is that only a portion of the genes that ever existed in the wild species was brought to cultivation in the processes of domestication, leaving most of the genes unused. With the help of molecular marker technology, it should therefore be possible to identify genes that can increase the yield of cultivated plants.

The second approach is to modify certain physiological processes by genetic engineering. Gan and

Amasino reported a system conceived to delay leaf senescence by autoregulated production of cytokinin. The construct was designed by fusing a senescence-specific promoter isolated from Arabidopsis with a DNA fragment from Agrobacterium encoding isopentenyl transferase (IPT), an enzyme that catalyzes the ratelimiting step in cytokinin biosynthesis. The strategy for such a system is that the gene would be turned on at the onset of senescence leading to the synthesis of cytokinin, and the production of cytokinin would in turn inhibit the process of senescence, thus repressing the expression of this construct itself.

Such a system would, therefore, be able to produce cytokinin for delaying senescence, and at the same time preventing overproduction of cytokinin, because overproduction of this hormone is detrimental to the plant. Transgenic tobacco plants carrying this construct showed a significant delay in leaf senescence, bringing about a large increase in the number of flowers, number of seeds, and biomass, indicating the possibility of increasing plant productivity by delaying leaf senescence. It is interesting, therefore, to determine if this system can provide a general strategy for yield increase in crop improvement. There are many opportunities for biotechnology to contribute to sustainable food production, to achieve higher yields, better quality, and less dependence on chemicals, making crop production more environmentally friendly.

Field Testing of Transgenic Crops

According to statistics from the Ministry of Agriculture, transgenic research has been conducted in 47 plant species in China using 103 genes. A national committee for the regulation of biosafety of genetically improved agricultural organisms was established in 1996 to promote biotechnology in a healthy environment. This committee accepts applications twice a year for biosafety evaluation of genetically improved agricultural organisms such as crop plants, farm animals, and microorganisms.

By mid 1998, the committee had received 86 applications, of which 75 were for field testing of transgenic crops. Permission for 53 of the applications was granted for commercial production, environmental release, or small-scale field testing. The crops used for transgenic research were rice, wheat, corn, cotton, tomato, pepper, potato, cucumber, papaya, and tobacco. A variety of traits were targeted for improvement including disease resistance, pest resistance, herbicide resistance, and quality improvement. In a few cases, transgenic crops have been grown for large-scale commercial production.

Intellectual Property Rights

One of the major constraints relates to intellectual property rights (IPR). China does not yet have effective IPR in place for large-scale biotechnology research to develop new genetically improved crops. Most of the genetically improved crop plants that have been developed so far involve complex IPR issues. There is a major shortage of experts in China with knowledge of IPR, and experience in dealing with these issues. China urgently needs help in training people in IPR. Scientists and breeders do not fully understand IPR, which are not always recognized and honored. Education is therefore urgently needed on these issues.

Delivery Systems

Another major constraint is the lack of delivery and extension mechanisms that take the products of biotechnology research to the farmers. China had a network system to dispense agricultural technologies, seeds, and other related materials. With the development of a market economy, the old distribution systems are gradually losing their effectiveness, and are now evolving into profitdriven seed companies undergoing the processes of privatization. Although this may be a good movement in itself, it may take several years for the system to become effective, because the funding situation does not appear to be promising at the moment. Governmental support mainly goes to the research component, and there is not enough funding to support initiatives and startups of seed companies.

Scientific and Technical Constraints

There are also a number of scientific and technical constraints to the application of technology in crop improvement. One of the constraints is the lack of understanding of the mechanisms governing the traits that are very important in crop improvement.

Drought causes severe yield loss worldwide, and it will continue to be among the most damaging stresses in crop production. Tolerance of the crop to drought as a trait, however, has not been well defined, and it is still not clear what aspects of plant morphology or physiology are the most important for drought tolerance. Research is still needed to define a clear target for improving drought tolerance.

There is also a huge need for germplasm. Germplasm has not been found for a number of important traits such as resistance to fungal diseases and resistance to a number of pests in crop species, for example, sheath blight of rice, scab disease of wheat, and yellow wilt of cotton. These have become the most devastating diseases worldwide, as have borer insects of a number of crops that cause heavy damage. International collaboration, coordinated by CGIAR centers, may have a crucial role to play in germplasm identification, exchange, and utilization.

Perspectives

Recent developments in genome mapping and genetic engineering have provided a knowledge base, identified germplasm resources, provided useful genes, and offered effective tools for crop improvement. Integration of the knowledge, the tools, and the genetic resources into breeding programmes will greatly increase the efficiency of new varietal development.

Molecular Marker-Assisted Selection

It is expected that molecular marker-assisted selection will have a major role to play in future genetic improvement of many crops. This is not only because the technique itself has provided a highly efficient tool for speedy and precise selection, but also because it possesses several distinct advantages. First, it does not require the isolation of the targeted gene, which often takes years and considerable resources to accomplish. Second, most of the gene constructs such as those commonly used in many transformation studies are now covered by IPR, hence are not freely available for varietal development. Third, the progeny developed by marker-assisted selection in general does not suffer from adverse effects such as over- or underexpression and transgene silencing, which are now frequently reported with transgenic plants. The performance of the progeny resulting from marker-assisted selection is therefore much more predictable than those from transformation. The large number of genes that have been precisely tagged and mapped will provide a rich source for marker-assisted breeding.

Gene Isolation

The most common practice for obtaining new genes is map-based cloning. Molecular markers that are closely linked to genes of interest can serve as the starting point for cloning the genes following the map-based cloning approach. It can be expected that the process of gene isolation using this approach will be greatly accelerated with advances of the international effort in DNA sequencing. It is highly likely that all the genes that are accurately mapped with closely linked markers can be quickly isolated with the availability of the sequence information.

The recent development in DNA-chip technologies may also provide a powerful tool for large-scale isolation of new genes in the near future. It can be expected that large numbers of genes will become available for crop improvement in the next decade. Biotechnology will soon play a major role in crop improvement in China. The area planted to cultivars, developed using modern biotechnology, will increase steadily in the years to come. Biotechnology will contribute significantly to food production and food security in China in the coming century.

9

Risk Assessment and Management

Agricultural biotechnology that have received widespread publicity in the media even though biotechnology has also been applied to health and industrial sectors. Environmental NGOs have been particularly vocal in taking issue with the new technologies derived from or incorporating GMOs. As a consequence, in the public debate biotechnology has become synonymous with GMOs, although they are only one of the many products of biotechnology. Curiously, biotechnology and GMOs in health-care products now in widespread use (insulin, hepatitis vaccine, medication for cardiovascular disease, etc.) or for industrial purposes such as bioremediation have elicited no such controversy.

This can probably be attributed to the lack of benefits to consumers in the first generation of genetically modified (GM) crops. The main focus was on herbicide and insect resistance that benefited farmers, seed producers, and chemical companies. It is expected that the next generation of genetically modified foods will benefit consumers, nutritionally or from taste or storage benefits, and accordingly may be better accepted.

A number of food-related crises in recent years have made consumers particularly sensitive about food safety issues. Health and food safety concerns are again at the forefront in Europe following additional cases of mad cow disease (bovine spongiform encephalopathy) and the banning throughout the European Union of blood and bone meal in feed for all animals. These crises have not been caused by GMOs, but by the intensification of agriculture and food production, a fact that appears to have escaped public attention. In Europe in particular, demands have been made for informative food labeling so that consumers may, if they wish, avoid genetically modified foods.

The anti-GMO movement reveals profound mistrust of developments in science and technology and of the forces seen to be driving them. Genetically modified crops such as maize, sorghum, cotton, and canola have been widely grown for the last five years, yet no harmful effects on human health or the environment have been detected. That was one of the conclusions of the OECD-sponsored conference held in Edinburg, UK, in 2000. However, it is generally agreed that government and the private sector are responsible for monitoring the long-term effects of GMOs on human health and the environment.

Potential Risks

The risks associated with modern biotechnology fall into four categories: food safety, environmental, socioeconomic, and ethical (Table 1). Some of these concerns relate to potential risks inherent in modern biotechnology and can be described as technology-inherent. Others are related more to value systems or cultural practices and can be described as technology-transcending.

Food Safety Risk

The potential risks of biotechnology on human health may

include toxic reactions, increased cancer risks, food allergies, food contamination, and antibiotic resistance (Table 1). There is also concern that GMOs in animal feed might present a health risk for consumers, or for the animal itself. Consumers are also concerned about the long-term health effects of genetically modified foods.

Table 1. Summary of Perceived and Genuine Risks of Genetically Engineered Foods and Crops

Nature of Risk	Type of Risk	Remarks
	1. Toxins and poison. In 1998, a scientist in the Rowett Institute found that GE potatoes spliced with DNA from the snowdrop plant (a viral promoter) are poisonous to mammals.	The UK Government's Advisory Committee for Novel Food and Process examined the data and concluded that the experiment was faulty and the conclusions were wrong.
	2. Increased cancer risks. Monsanto's bovine somatotrophin (growth hormone) injected into dairy cows to produce more milk has been reported to cause cancer in human breast, prostate, and colon.	This is not a GM food. In any event, Canada and the European Union have banned its use. A United Nations Food Standard body has not certified its safe use.
	3. Food allergies. In 1996, a Brazil nut gene spliced into soybean was reported to induce potentially fatal allergies in people sensitive to Brazil nuts.	The hormone is no longer widely used in US. The safety assessment confirmed that the protein was an allergen and the development was abandoned. A standard laboratory test has been available to test possible allergenicity in GE products.

	4. Contamination. StarLink, a GE maize variety approved for animal feed but not for human consumption, was found in an ingredient used by some US beer makers and in taco shells in the US in 2000.	The incident was caused by an accidental mix of StarLink with vast amounts of other maize during harvest, storage, and distribution. The contaminated food was recalled and destroyed. A number of quick and cheap tests are available to determine the presence of GM products in food.
	5. Antibiotic Resistance. Use of an antibiotic marker gene in the development of GE crops may contribute to the growing public health danger of antibiotic resistance.	There is little or no evidence about this risk yet. But this is an emotive topic, and developers have now replaced the antibiotic marker with a safer marker.
Environmental risk	1. Increased pesticide residues. Farmers growing GE crops will use as many toxic insecticides and herbicides as conventional farmers, thus increasing pesticide residues in soils and on crops.	This risk is not yet proven statistically. There are reports that farmers growing GE crops resistant to pests and herbicides are able to reduce production cost significantly through the reduced use of pesticides. That was a major reason why farmers adopted GE crops widely in the PRC and the US.
	2. Genetic pollution. Wind, rain, birds, and bees have carried genetically altered pollen into adjoining fields, contaminating the DNA of organic, non-GE crops.	This genetic pollution is not an environmental issue unless the transfer of pollens causes some kind of environmental damage. Pollen contamination has taken place for centuries with

	or without genetic engineering.
3. Damage to beneficial insects. Scientists from Cornell University found that pollen from Bt maize was poisonous to Monarch butterflies and may be to other beneficial insects as well.	Monitoring systems have been devised in the PRC and the US to evaluate the long-term effect of GE crops on beneficial insects.
4. Creation of superweeds. GE crops (soybean and canola) resistant to herbicides may transfer their resistance to weeds, turning them into superweeds, which cannot be controlled by herbicides.	This fear has yet to be proven. Scientists are closely monitoring the use of GE crops resistant to herbicides.
5. Creation of superpests. GE crops (maize and cotton) resistant to pests may transfer their resistance to pests, turning them into superpests which cannot be controlled by pesticides.	As above, this fear has yet to be proven in practice. There is no known mechanism by which pest resistance from a plant may be transferred to an insect pest.
6. Creation of new viruses and bacteria. Biotechnology could help terrorists to create killer viruses or bacteria, which could be used in biological weapons.	This could happen, even without biotechnology. Terrorists historically have managed to acquire and subvert beneficial technologies to antisocial purposes.
7. Genetic bioinvasion. By virtue of their superior genes, some GE plants and animals will inevitably run amok,	There is as yet no scientific evidence that such plants and animals can be created through biotechnology.

	overpowering wild species in the same way that introduced exotic species do.	
Socio-economic Risks	1. Terminator technology will render seeds infertile and force hundreds of millions of farmers to purchase more expensive GE seeds and chemical inputs from a handful of global biotechnology and seed companies.	The Monsanto Company has withdrawn the terminator gene from its GE crops following many complaints from farmers.
	2. High concentration of biotechnology research and development in developed countries will widen the income disparity between developed and developing countries, and between large and small farmers.	The public sector in Asia should accord high priority to biotechnology development that address the problems of small farmers.
Ethical Concerns	1. Biotechnology reduces all life to bits of information (genetic code) that can be rearranged at whim by scientists. The creation of the first genetically modified monkey in 2000 brings the possibility of genetic manipulation closer to humans. There is fear that the technique will be used to create "designer babies." 2. There seems to be little ethical concern by the private companies over the use of GE animals to produce therapeutic drugs.	Although most of these ethical concerns relate to non-agricultural biotechnology, they point to the need for the private sector to incorporate work ethics in biotechnology research and development.

To address food safety concerns, the following safeguards have been adopted by some countries:

(i) Some countries have regulatory procedures, institutions, and infrastructure in place to ensure food safety. These regulations cover all aspects of the food chain, from farm inputs (including animal feed, feed additives, pesticides, fertilizers, veterinary drugs) through production and processing (including agricultural products, processed food, novel foods, food additives), to transportation, storage, and distribution.

(ii) Formal science-based procedures for risk analysis of food have been adopted by some countries. These continue to evolve with new scientific information about food safety, emerging pathogens, new technology, and consumer demands for a high level of public health protection. Generally, the procedures conform to international standards set by the Codex Alimentarius Commission of FAO.

(iii) Some countries have adopted a variety of approaches to regulate genetically modified foods, either by applying existing food safety measures or enacting new legislation.

(iv) In response to consumer demands, a growing number of countries are introducing labeling. Opinions differ on whether labeling should be mandatory or voluntary, as well as on acceptable tolerance levels and on the type of information to be used. Two labeling issues currently being addressed are segregation and traceability. Segregation relates to the ability to attest to the separation of genetically modified and non-genetically modified crops (as in the case of soybean in processed foods). Traceability means being able to attest to the origins of food products. Traceability is of particular importance in the current controversy related to mad cow disease. Low cost, quick tests are now available to determine the presence of genetically modified products in food.

Environmental Risk

A number of biotechnology applications are not seen to present any new threats to the environment. That is the case with tissue culture, diagnostics, and market-selected plant breeding. On the other hand, there is fear of potential risks from the release of GMOs into the environment. The potential risks of GMOs on the environment may include increased pesticide residues, genetic pollution, damage to beneficial insects, creation of superweeds and superpests, creation of new viruses and bacteria, and genetic bioinvasion (Table 1).

To address environmental concerns, some countries have adopted the following safeguards:

(i) Most current biotechnology applications in tissue culture and micropropagation, diagnostics and vaccines, and markerassisted plant breeding are subject to existing regulations. They include phytosanitary regulations and plant quarantine, varietal certification of seeds, and veterinary product regulations. These regulations usually conform to international standards, guidelines, or recommendations such as those set by the International Plant Protection Convention or the International Bureau of Epizootics for animal products.

(ii) A number of countries have introduced new requirements and procedures for the environmental release of GMOs. Procedures for hazard identification and risk assessment of GMOs are now well-established in most OECD countries, and in some Asian countries.

Socioeconomic Risk

Modern biotechnology R&D has been conducted in an institutional and economic environment that differs significantly from the development of the earlier Green Revolution technologies. While the latter were essentially the prerogative of public research institutions and philanthropic foundations, developments in biotechnology have been driven essentially as a competitive, commercial endeavor in which powerful private sector actors compete.

The major socioeconomic risk of agricultural biotechnology stems from the fact that the research, development, commercialization, and distribution of new biotechnological products have been carried out mainly in developed countries by a few, large, multinational companies. These companies have focused on temperate crops for large farmers in developed countries. Undertaking R&D on Asia's basic food crops for small farmers in rainfed and marginal areas is of little interest because they see limited returns from such investments.

If this trend continues, modern biotechnology will aggravate the income disparity between developed and developing countries, and between large and small farmers. Increased public investments in agricultural biotechnology are necessary to ensure that small, poor farmers have access to biotechnology. Governments must also address the potential gender disparity in the access to technology, and the negative impact specific to women, since no technology is gender neutral. Lessons from earlier agricultural technological changes should be used as a caveat.

Ethical Issues

Ethical issues may stem from the uneven or inequitable impact of an expanding global economy; from the national social, economic, and politi-cal context; or from individual values. Indeed, it may be difficult to reconcile personal values with what a majority regard as the common good. One of the ethical concerns raised by biotechnology-and particularly GMOs-is that it is unnatural and an unwarranted tampering with nature. However, seen in historical perspective, most technology developments in agriculture over the centuries have involved, in one way or another, efforts to overcome the vagaries of nature.

Risk Management

To address the potential risks of biotechnology in agriculture, there is a need for Asian countries to establish effective biosafety procedures for the development, testing, and release of the technologies. An example on how this should be done is shown in Figure 1.

A growing number of international organizations have become involved in food and environmental safety. Concerted international efforts to develop agreed upon scientific approaches to biosafety date from 1975, when a group of scientists from the United States National Academy of Sciences expressed concern regarding the potential biological risks of rDNA molecules. Efforts to ensure the science-based, case-by-case risk identification, assessment, and management of GMOs have thus been pursued for some 25 years.

In the industrialized world, OECD has been instrumental in forging agreements on scientific principles regarding the safe applications of rDNA organisms in industry, agriculture, and the environment. Since the 1980s, OECD has established guidelines, first for (i) laboratory-based experimentation, then (ii) for small- and large-scale field trials of genetically modified plants and organisms, and finally (iii) for their commercialization and release into the environment or food chain.

During the 1990s, a growing number of international organizations including the United Nations Environmental Programme (UNEP), UNIDO, World Health Organization (WHO), and FAO have become involved in activities related to biosafety. Both UNIDO and UNEP have developed biosafety guidelines. FAO and the WHO are jointly responsible for the Codex Alimentarius Commission, which sets international standards for food safety.

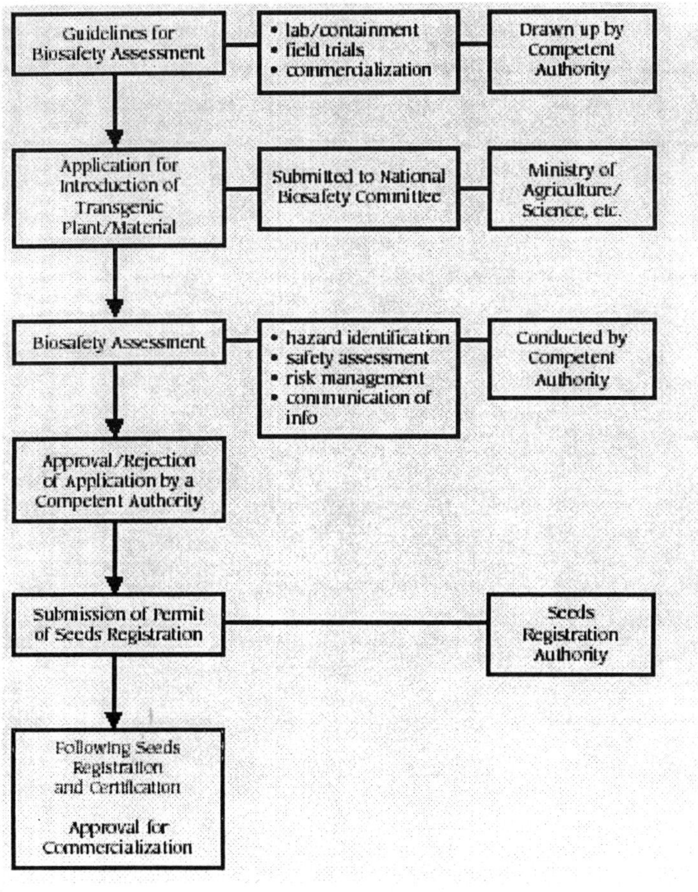

Figure 1. Procedures for Release/Commercialization

Codex has set up a new Intergovernmental Task Force on Foods Derived from Biotechnology to develop general principles for risk analysis and to provide guidance on risk assessment. Other international organizations conducting work on the safe use of biotechnology include the International Centre for Genetic Engineering and Biotechnology, the United Nations Conference on Trade and Development, and, for many aspects of animal as well as human health, the International Bureau of Epizootics.

Preoccupation with the protection of biological diversity, necessary for sustaining agriculture and food production and, indeed, life itself, became a topic of international debate during the 1980s. As a result, FAO established the International Undertaking on Plant Genetic Resources as a nonlegally-binding agreement for cooperation in the conservation of genetic material. The agreement was based on the universally accepted principle that plant genetic resources are a heritage of mankind and consequently should be available without restriction.

The legally binding Convention on Biological Diversity, which came into force in 1993, encompasses not only plant genetic resources but all living organisms. In contrast to the FAO undertaking, the Convention affirms that States have sovereign rights over their own biological resources. In January 2000, the legally binding Cartagena Protocol on Biosafety was adopted. This Protocol lays the foundation for a global system for assessing the impact of GMOs on biodiversity. It outlines the obligations of all countries that are signatories to protect biological diversity from the potential risks imposed by living modified organisms resulting from modern biotechnology.

At the heart of the Protocol is the concept of advanced informed agreement to the import of GMOs. This means that an exporting country must inform an importing country of its intention to export. It must also provide information on the GMOs, including an appropriate risk analysis. For its part, an importing country is required to give its informed agreement to the import. This implies that, although signatory states have the right to prohibit the import of GMOs if advanced informed agreement has not been given, they must also be capable of determining the validity of a safety assessment.

In addition, the Protocol advocates the "precautionary approach," described in the 1992 Rio Declaration on Environment and Development as a principle of international environmental law. The Protocol also proposes to establish a Biosafety Clearing-House for collecting, sharing, and disseminating information on risk assessment and management, which is particularly important for developing countries. Thus far, 81 countries have signed the Protocol, including Bangladesh, PRC, Indonesia, Republic of Korea, Malaysia, Philippines, and Sri Lanka.

A sample of industrialized countries shows a diversity of approaches to food and environmental safety and in the agencies or ministries that play key roles in the national biosafety system. For example, the US has chosen to develop biosafety capacity within existing institutions (notably the Environmental Protection Agency and the Department of Agriculture). Denmark has designated the National Forest and Nature Agency as the lead agency for biosafety. In Norway, in accordance with the Gene Technology Act, the application of GMOs is subject to ethical and social conditions.

A GMO can be approved for use only after the assessment of its benefit to the community and its contribution to sustainable development. In Australia, apart from the different ministries involved, an Advisory Committee on Genetic Engineering was created some years ago. That will soon be replaced by an office with an enforcement rather than an advisory role. A public hearing must be held before authorization is given to release a GMO. Australia has recently created the Office of the Gene Technology Regulator to operate in the Commonwealth Department of Health and Aged Care.

Japan has based its risk assessment system on the principles and concepts developed by OECD. Different Japanese ministries and agencies have developed guidelines in accordance with their responsibilities on a product-sector basis. A growing number of countries are introducing guidelines or mandatory measures for labeling genetically modified foods. In the US, formerly strongly opposed to labeling, the Food and Drug Administration has recently been charged with developing guidelines for voluntary labeling of food products to indicate whether or not they contain GMOs.

In the European Union, food products containing more than 1 percent GMO must be labeled, stating clearly: "this product contains GMOs." The Australia-New Zealand Food Standards Council is drawing up a new Food Standards Code. Labeling requirements will be drawn up in accordance with this joint Code. Japan is planning the introduction of mandatory food labeling in April 2001. The OECD has also played an active role in promoting the harmonization of biosafety procedures among its member countries so that information and data gathered in the course of risk assessment may be mutually acceptable.

Regulatory Systems

Most countries have not yet completed guidelines for commercialization. The PRC is the only country in Asia where transgenic crops have been approved and released for sale. A number of countries are also in the process of introducing food labeling. Indonesia is planning to introduce mandatory labeling in April 2001.

ASEAN has undertaken a number of initiatives aimed at the harmonization of biosafety procedures among its member countries. In 2000 it also conducted a joint workshop with the Ministry of Science and Technology of the PRC to discuss issues related to transgenic plants. The Asia-Pacific Economic Cooperation has a number of capacitybuilding and technical cooperation activities in support of risk assessment, risk management, risk communication, and public acceptance. Responsibility rests with its Agricultural Technical Cooperation Experts Group and its Sub-Group on Research, Development and Extension of Agricultural Biotechnology.

Although some Asian countries have regulations in place, weak law enforcement sometimes results in the release of GMOs without proper procedures having been followed. Law enforcement in Asia needs strengthening to avoid risk to human and environmental health.

International and Donor Support for Biosafety Capacity Building

A growing number of initiatives are being taken to provide support for Asian countries in developing their capacities to safely manage biotechnology. Following the signing of the Biosafety Protocol, UNEP, through its Global Environment Fund, is preparing a major capacity-building program for developing countries. It will also be responsible for the establishment of the Biosafety Clearing-House.

FAO serves as the secretariat for the International Plant Protection Convention, which has recently established a working group on the Phytosanitary Aspects of GMOs, Biosafety and Invasive Species. UNIDO has conducted training courses in biosafety in a number of Asian countries. In collaboration with OECD, it has compiled an information source and database (BIOBIN), which provides information on biosafety methods and procedures, and on the status of the regulatory frameworks in individual countries.

OECD has established a network of international organizations with activities related to biosafety to enhance the exchange of information and to facilitate cooperation.

IBS provides a number of services to enhance developing country capacities to manage agricultural biotechnology. It gives short training courses annually for managers of agricultural biotechnology research programs in Asia. Biosafety is a key element in these courses. IBS also provides advice to individual countries.

USAID is involved in a number of biosafety training activities in Asian countries, some of which have been provided by its Agricultural Biotechnology Services Program.

ISAAA is currently setting up a Global Knowledge Center on Crop Biotechnology based at ISAAA's SEAsia Center at Los Baños in the Philippines. ISAAA also organizes capacity-building workshops and training for regulatory officials and scientists from the Asian region.

Genetically Modified Organisms

In industrialized countries, considerable progress has been made in methods, approaches, and experience in the safe management of GMOs. Overall these methods have proven to be effective, particularly with respect to food safety. The development of a transgenic soybean containing a protein derived from Brazil nuts, potentially useful in animal feed, was abandoned when the safety assessment revealed that the protein was probably an allergen. In response to consumer concerns, antibiotic markers are now being phased out as alternatives are being developed. That despite consensus among national and international regulatory authorities and the scientific community that the markers pose no threat to human health.

Despite acknowledged progress, areas of scientific uncertainty or disagreement persist. New challenges for risk/safety assessment, management, and monitoring will arise as a second generation of food, agricultural, and public health products emerges during this new decade. Not only are the ranges of organisms and numbers of traits expected to expand, so are the numbers and diversity of geographic and ecological sites in which they are released.

These new products will not lend themselves easily to current approaches and methods to risk and safety assessment and to management. This highlights the need for constant review and improvement of assessment principles and procedures, and the urgent need for collection and analysis of ecological data. Other areas where consensus has not been reached among scientists themselves, or among scientists and policymakers, include (i) the precautionary approach as a method of dealing with scientific uncertainty, and (ii) methods for traceability.

The potential for the development of insect resistance to Bt transgenic crops is another area where the scientific community, industry, and environmental NGOs have not reached consensus. Risk management approaches at the farm level have so far been applied mainly on large-scale, commercial farms where monoculture is practiced. The same approaches may not be effective and, indeed, may not be necessary in the mixed farming systems of Asia.

Efforts to address the issue of resistance to transgenic crops have recently been initiated in Asia. A first International Consultative Workshop on Effective and Sustainable Use of Agricultural Biotechnology in Integrated Pest Management in Developing Countries was held in November 2000. Organized by IRRI, Zhejiang University, and the International Organization for Biological Control, the workshop tried to determine strategies for the safe deployment of crops such as cotton and rice with novel Bt genes for pest resistance.

Among OECD countries, the continued monitoring of GMOs after release into the natural environment is also an area where opinions differ. On the one hand, it is argued that monitoring should be continued so that unforeseen risk can be managed and ecological impacts can be assessed. On the other hand, it can be argued that the high costs of monitoring are not justified when transgenics have already successfully passed the hurdles of risk identification, safety assessment, and risk management. It is clear that the potential risks of GMOs may vary by site and over time. Continued research will be necessary to improve data collection and analysis and to modify or devise new risk management strategies as needed.

Future Perspective

To date, a number of Asian countries have made substantial investments in biotechnology R&D initiatives. Some public research institutions have reached the stage of field testing of new crop varieties, major research projects are coming to fruition in IARCs in the region, and private companies are also conducting research in the region, or have products waiting to be commercialized and exported to the region. It will be necessary to have biosafety procedures in place to ensure that the benefits of these technologies are realized.

Although many opportunities exist, from donor and other development agencies, in support of capacity building for the safe management of biotechnology, and in setting up national systems, it is the responsibility of national governments to ensure that national regulatory systems are applied, enforced, and monitored. Setting up an effective national regulatory system for the safe management of biotechnology requires substantial resources, both financial and human. It is also a political process and, as such, requires political commitment.

A country is free not to allow the import of GMOs on its territory. At the same time, it has an obligation to establish a national biosafety system. This means that even if a country decides to ban the import or use of GMOs on its national territory, it will still need to comply with the obligation to develop a biosafety system for the control of such organisms. It is anticipated that UNEP, the World Bank, and United Nations Development Programme (UNDP), through the Global Environment Fund, will contribute funding and technical assistance to developing countries for capacity building in biosafety. Even so, compliance with both the provisions of the Convention on Biological Diversity and the Cartagena Biosafety Protocol will place a heavy burden on some countries.

Apart from obligations under the Biosafety Protocol, if countries wish to control the import or introduction of GMOs on national territory, it will be necessary to have the means of detection in place. Similarly, from the perspective of international trade, when countries export agricultural or food products to countries that either prohibit the import of GMOs or require mandatory labeling for food or food products containing GMOs, the requisite regulatory procedures need to be in place.

The regulatory framework for biotechnology should not, however, be regarded in isolation from the broader policy context of agriculture and the contribution that biotechnology might make in the particular economic, social and environmental context of individual countries. Countries have a range of options in determining policies and priorities for biotechnology. These might range between active promotion of biotechnology approaches, "wait and see," or rejection. Similarly, countries have a range of options in designing and implementing a national biosafety system geared to their particular technical, legal, and institutional realities. For example, a complex and time consuming regulatory system might impose prohibitive costs for local enterprises, particularly start-up companies.

China is the only Asian country so far with experience of managing and monitoring new transgenic crops in the field on a large scale. Some countries are gaining experience in field testing, while others have not yet completed biosafety guidelines. Information on socioeconomic and scientific developments in biotechnology and biosafety can be accessed from a number of information sources.

10

Biotechnology to Sustainable Agriculture

Agricultural production systems have real resource limitations. There is limited land area suitable for agricultural production, which land resource scientists believe will see a net increase of 5% over the next century. Other limitations include fossil fuels-most temperature zone crops require more fossil fuel for production than the plants capture as harvestable photosynthate. Based on current sources and rates of consumption, we also have limited supplies of phosphorous and potassium. A second, perhaps equally important concern is the instability of agricultural production environments.

Predictions of climatic change suggest that global mean temperatures could change by either plus or minus 2-4°C. While it is unlikely that we will experience such radical mean temperature changes over the next decade, it is realistic to expect some temperature and associated precipitation changes over the next 20 to 50 years. Net changes in mean temperature over the next 50 years will probably reflect the balance between increasing greenhouse gases and changes in atmospheric particulates from volcanic activity and combustion. Desertification processes appear active in both Africa and Asia. Clearly, climate is unstable, and may have an impact on the stability of agricultural production over the next century.

If we examine total world agricultural production over the past 30 years, two dramatic changes are perceptible. First, agricultural production has increased dramatically (on par with population growth). Secondly, there is increasing annual variation in total production around that trend line. Variation has increased substantially in the past few decades, perhaps as a result of the natural interplay between variable climatic conditions and increasingly intense agricultural output.

A third major source of uncertainty is associated with variable economic and trade policies, including uncertainties associated with the implementation and effect of the recently concluded negotiations of the General Agreement on Tariffs and Trade (GATT). Mr. Arthur Dunkel, former Director General of GATT, proposed the process in the hope that GATT would eliminate agricultural production subsidies, and there will certainly be an impact, whether positive or negative. All three areas of uncertainty (resources, climatic change, and trade policy) will determine when-or if-the global agri-food system will be sustainable.

What impact can biotechnology have on agricultural sustainability or global food security? Biotechnology may help in global agricultural sustainability if it accelerates genetic advances in productivity. There has been much talk about biotechnology in a general way, in some very specific applications. Increases in productivity in the past 100 years-which have been huge and have accommodated a 400% increase in human population-have relied on two major components: genetic advances and increases in resource inputs.

Many of our agricultural resources are limited, so in the long term it is very important to use resources wisely.

Future sustainability will probably also come about through genetic advance. The maize industry in North America and Europe has increased productivity at a rate of two bushels per acre per year in recent decades, mostly through improved hybrids. Similar relative gains through genetic advances have been made in cereals, soybeans and most other major grain crops. Gains in vegetable crop productivity have been somewhat slower over that period.

If biotechnology can help us advance the genetics of our food crops (greater output per unit of finite resource input), then it can contribute towards agricultural sustainability. If biotechnology can help preserve environmental resources, and both essential resources (air, water and nutrients) and also aesthetic resources (green space, parks, diversity, etc.), then biotechnology may help us move towards agricultural sustainability.

If biotechnology (and plant breeding and genetics) fails to help improve resource utilization efficiency, then agriculture will probably not be sustainable, and the size of the global community will decline as resources are exhausted. If biotechnology fails to help preserve aesthetic components of the environment, then there will also be a decline in the quality of life for the global population.

Environmental considerations are critical to sustainable agriculture, and biotechnology may contribute to enhancing or conserving environmental resources, if properly managed. If not properly managed, biotechnology could accelerate environmental degradation.

If biotechnology is really to have an impact on sustainability, real sustainability, then it has to be available to those who need it. Certainly, the pilot programmes of the International Service for the Acquisition of Agri-biotech Applications (ISAAA) are an effort towards that end. But there are many additional efforts that are required. One area often ignored, in the context of technology transfer, is human resource development in those regions that can best benefit from biotechnology, in order to have the expertise available and apply it in the most appropriate ways to answer regional needs. Technology transfer is an essential requirement if biotechnology is going to have a real impact on global agricultural sustainability.

The real challenge has been an evolution of appropriate regulatory processes that ensure biotechnology impacts positively on the major factors related to agricultural sustainability, while protecting environmental resources and human health and welfare. If, through regulatory processes, we can move towards positive benefits from biotechnology, sustainability will be one step closer-that is, sustainability may become a real possibility, rather than the major uncertainty it is today.

Ecological and Evolutionary Problems in Transgenic Crops

The genetic improvement of cultivated plants through the use of recombinant DNA techniques promises to enhance the quality of food and fiber, to increase yield through better protection from diseases and pests, and to reduce the need for chemical inputs-both toxins and fertilizers. Because many of these advances allow more efficient use of land already in agricultural use, they promise not only to feed and clothe people but to help alleviate the pressure for conversion of wildlands into croplands and pastures. The development of improved tree varieties for reforestation of degraded tropical agricultural lands, through rDNA techniques, could provide a sustainable supply of forest products that would further relieve pressure on tropical forests.

What, then, are the conceivable ecological hazards associated with the testing and use of transgenic plants? There are six categories of environmental concern:

1) the creation of new weeds;
2) the amplification of the effects of existing weeds;
3) harm to nontarget species;
4) disruptive effects on biotic communities;
5) adverse effects on ecosystem processes; and
6) squandering of valuable biological resources.

Creation of New Weeds

Many crops have been domesticated to such a degree that they are entirely dependent on human activities. Maize, wheat and bananas are good examples. Such crops, whatever the improvements made through biotechnology, are unlikely to become self-propagating weeds in any ecological context. Some other cultivated plants are weeds in some contexts and crops in others-certain kinds of millets (Pennisetum spp.) and sorghums (Sorghum spp.) fall into this category.

The introduction of a gene that increases plant fitness (such as resistance to disease or to pests) to a crop in this category might shift the balance toward weediness in areas where they are now safely grown as crops, or promote weediness in varieties now considered safe. Other crops have close weedy relatives-including sugarcane (Saccharum), rice (Oryza), potatoes (Solanum), sweet potato (Ipomoea), vegetable and oil seed (Brassica), sunflower (Helianthus), and oats (Avena).

For some of these crops, which still share many characteristics with their weedy ancestors, certain kinds of genetic alterations might create weed problems with the crop itself. For example, a transgenic, highly salt-tolerant variety of paddy rice might itself invade estuaries. Generally, however, the likelihood of most crops themselves becoming serious weeds is small.

Considerably more likely is the development of increased problems with already-weedy relatives, by the acquisition of fitness-conferring genes from a transgenic crop through hybridization and introgression. Ecologists consider this scenario to be the primary ecological risk from transgenic plants. It is often wrongly discounted, however, due to a common confusion between the concepts of introgression and hybridization. Some pairs or groups of closely-related plant species (for example, orchids or cucurbits) hybridize freely in cultivation, free from the constraints of the co-evolved pollinator systems of their ancestral habitats. For other groups of plants, hybridization is infrequent, rare, or completely impossible. If formed, first-generation (50/50) hybrids may be fully fertile, but more often they have reduced fertility, or are completely sterile.

Introgression is the incorporation of genes from one species into the gene pool of another. The process must begin with hybridization, but it is more complex. Introgression in nature is precisely analogous to a common technique of plant varietal improvement. The plant breeder may start with a variety that has many desirable commercial characteristics, but is susceptible to some pathogen. The breeder crosses this variety with a wild relative that lacks the commercial traits, but is resistant to the pathogen. Often, the first-generation hybrids that have the resistance trait will have lost many of the desirable commercial traits of the cultivated parent. These must be recaptured by repeated backcrossing and selection.

The process of hybridization plus backcrossing and selection is an exact parallel, under artificial selection, to the process of introgression under natural selection. Just as the breeder's first generation hybrid lacks the combination of traits that makes it either a fit wild plant or a desirable cultivated plant, the first generation hybrid between a transgenic crop and a weedy wild relative at the edge of a test plot or a farmer's field would likely be rejected both by the farmer and by nature-its fitness relative to the wild plant would be low.

But if it flowers and passes pollen to its wild cousins, the next generation will have ¾ of the genes of the wild relative, the next 7/8, and so on: this is the process of introgression. If the hybrid included a transgenic trait that would, by itself, confer high fitness on the weedy relative such as disease or pest resistance-then natural selection will rapidly promote the reproductive success of these successively "wilder" genotypes, just as the plant breeder regains the commercial qualities of the new variety by backcrossing.

Hybridization need not be a common event for introgression to proceed, carrying a transgenic trait from crop to weedy relative. That a crop and a weedy relative are known to hybridize only "rarely" is sufficient for the escape of a transgenic trait into the population of a weedy relative. If the gene does as much for the weed as the genetic engineer hopes it will do for the crop, then, at the limit, hybridization need only occur once for escape; natural selection will take care of the spread of the gene. Likewise, even crops such as sugarcane (Saccharum officinarum) and sweet potato (Ipomoea batatas), which are generally considered infertile, occasionally produce normal pollen; if hybrids form, once may be enough.

The second point is that hybrids need not be particularly fit in themselves, as long as they are competent to backcross with the weedy relative. Maize (Zea mays) and its wild relatives, the teosintes of Mexico and Guatemala, provide a good example. Teosintes include three wild subspecies of maize itself (Zea mays mexicana, Z. mays parviglumis, and Z. mays huehuetenangensis) and three closely related species of Zea (Z. dipolperennis, Z. perennis and Z. luxurians).

Some of these taxa are agricultural weeds and some are wildland species. Hybrids between teosintes and cultivars of maize are known for several of these taxa. Supposedly, some Mexican farmers even encourage some hybridization by allowing weedy teosintes to grow at the edge of their fields to make the maize "stronger," but genes may flow from maize to teosinte in such cases, too. In all cases, the hybrid is of low fitness, as evaluated by both natural selection (the seeds do not disperse easily) and artificial selection (to the farmer, the grain is inferior in quality and is not as easily harvested as maize, so is not kept for seed).

Nonetheless, introgression from cultivated maize into wild teosinte populations. Such introgression simply indicates that, however low the fitness of first-generation hybrids, they are nonetheless competent to backcross with teosinte. Because of the low fitness of intermediates, maize and the teosintes have kept their distinctive identities; an evolutionary biologist might call it a case of disruptive selection. Some cite the fact that maize and teosinte have coexisted for centuries without losing their identities as evidence against the danger of creating a seriously weedy teosinte through acquisition of a transgenic trait, and against the danger of genetic contamination of wild teosinte gene pools with novel genes of distant origin (such as the Bacillus thuringiensis gene-B.t.).

These reassurances, based on a misunderstanding of the nature of introgression, are entirely misguided. It is also well to remember that when low-probability events (or hazards) are multiplied by large exposures, they become virtually inevitable. Even if the outcrossing rate of a crop plant is only 0.1%, if there are a million flowers in field, then we must expect 10,000 outcrossing events. The irrelevance of traditional plant quarantine regulations and practices, with respect to guarding against introgression of transgenic traits into wild or weedy relatives of cultivated plants. Because a gene is not a disease or a pest, quarantine is pointless. The danger point is the open field somewhere in the middle of the countryside, not a port of entry.

Unfortunately, small numbers of familiar local weeds that contain a newly acquired, high-fitness, transgenic trait are likely to go unnoticed, simply because they are familiar, until they become a significant problem. Thus, awareness of the potential for introgression and vigilance regarding potential recipients of such genes are called for not at the port of entry, but around test plots and, later, in the first commercial fields. The exception would be the very difficult task of controlling, at ports of entry, the illegal importation of small quantities of transgenic seed by individuals, even well-meaning farmers or amateur gardeners. Again, as we know from countless cases of "informal" importation of weeds and other undesirable species by individuals, one mistake can be costly.

In the case of transgenic plants, the nontarget species of concern include both animals and other plants. When genes producing compounds intended to deter or kill pest insects or mites are incorporated into crop plants, the ideal result is that only the target pests are affected. We know from decades of unhappy experience that this ideal result is all too often not the case with chemical pesticides. Of course, our great hope is that transgenic plants with highly specific secondary chemicals are the answer.

Precautions are needed, however. An insect-pollinated transgenic plant that has been engineered to make an insecticidal compound intended to deter leaf or root-eating pests had better not produce the same compound in its pollen or nectar. A fiber plant that produces an insect toxin in its leaves had better not poison the farmer's cattle that wander into the field. A plant that produces a mite toxin would likely affect beneficial predatory mites as well as phytophagous mites.

Even B.t. toxin, of which some strains affect only lepidoptera insects, could potentially harm beneficial-or at least desirable-butterflies and moths. For example, the striking red-, orange-, or yellow-and-black heliconine butterflies of tropical forests not only help attract many naturalhistory tourists to tropical countries, but are considered to be "keystone species" in most natural communities in the New World tropics. The larvae of most of this diverse group of butterflies feed on wild species of Passiflora or passionfruit vines (maracuyá). Because certain lepidopteran larvae are also pests of cultivated passionfruit, incorporation of the B.t. gene would seem a natural approach to improving this crop.

The wildland butterfly species would be severely affected, however, by the acquisition of the B.t. gene by these wild plants through introgression with a cultivated transgenic passionfruit carrying the gene. Thus, unless hybridization can be prevented, it would seem dangerous to field test or commercialize passionfruit cultivars with the B.t. gene anywhere near wild Passiflora spp. Harm to other plants caused by transgenic crops includes several potential problems, all of them consequences of the acquisition of transgenic traits by wild relatives through introgression.

Effects on Biotic Communities and Ecosystem Processes

The composition and relative abundance of species, and the spatial structure of natural plant communities, depends upon a complex balance between plant-plant competition, the effects of herbivores and seed predators, and interaction with pollinators, seed dispersers, and soil mutualists. The acquisition of a high-fitness trait, such as protection against herbivorous insects, by a wildland plant species through introgression with a related transgenic crop could have several disruptive consequences.

We know from agricultural experience that pests are capable of causing massive reductions in crop reproductive fitness and yield. Experiments in natural communities and the record of successful biological control of weeds by imported insects and pathogens testify to the importance of reproductive control of plants by enemies in both agricultural and natural communities.

A wildland plant species, released from significant natural control, would become a better competitor, likely to reduce the density of competing plant species. Secondary effects could include declines in animal populations dependent on these species, and even changes in vegetation structure.

Ecologists have expressed concern about the effects of transgenic plants, or wild plants that acquire transgenic traits through introgression, on the functioning of ecosystems. Of course, transgenic trees used in reforestation projects may have very beneficial effects on ecosystem processes by increasing the per-area carbon fixation rate, by stabilizing or enriching soil, and by buffering climate.

But some caution is called for regarding transgenic (or other) plants that significantly affect soil nutrient balance. For example, if and when nitrogen fixation gene complexes are successfully introduced to forest trees, careful testing should precede widespread introduction in order to determine the effects on the supply and demand for other soil nutrients, particularly phosphorus.

Squandering of Valuable Biological Resources

The principle biological resources that may be at risk from the testing and use of transgenic plants are plant genetic resources-in this case, not so much the genetic diversity of land races and cultivars, but diversity of the gene pools of wild species. Once again, the principal risk arises from introgression of cultivated crops with wild plants.

Some, such as African oil palm and cacao, are important components of the Costa Rican agricultural economy. In the case, at least, of oil palm, Costa Rican wild relatives have already been the source of genetic variation for past crop improvement.

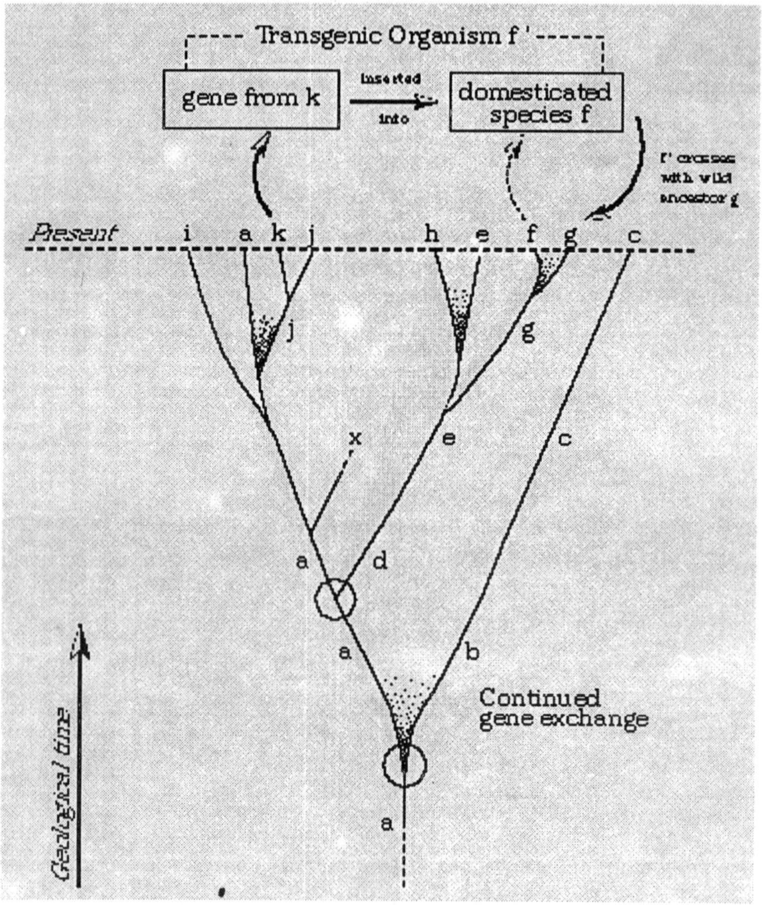

Figure 1. The "Conduit" Effect

Indeed, an ethical argument can be made against allowing introduction of genes from unrelated organisms to wild organisms of any species. For example, we know that genes from cultivated maize have entered the gene pools of wild teosintes. This can be considered a natural process that is part of the evolution of crops from wild relatives. But the introduction of, for example, a gene from bacteria (B.t. toxin), or from a distantly related plant group, seems to some of us to be a different matter-a process referred to as the "conduit effect" (Figure 1).

There is no adverse ecological effect, such genetic contamination devalues the genetic resource and evolutionary integrity of wild species. This ethical argument, of course, rests on the distinction between domesticated species, which have long evolved within the evolutionary domain of humans, and wild species, which have historically been comparatively free from our influence. Such congeneric matches are simply the most likely suspects for introgression-suspects that a wise regulatory policy should insist on knowing more about before granting permission for even an initial field trial, in my opinion.

One final resource that transgenic plants may threaten, if used unwisely, is the very small library of highly specific genes for the control of insect pests; the most obvious example is B.t. As for any other pesticide, resistance can evolve quickly if pest populations are continuously exposed to it over large areas, eliminating the pockets of susceptible genotypes whose resurgence is the key to continued effectiveness.

At present, a special cooperative effort between ecologists, evolutionary biologists and molecular biologists is underway, funded by the biotechnology industry, to find ways to prolong the useful life of the B.t. gene by minimizing exposure and pinpointing expression to vulnerable tissues, instead of relying on constitutive expression of the trait. This cooperative project is a fitting example to end with, showing the potential role that collaboration between ecologists and evolutionary biologists may play in the safe and effective implementation of biotechnology.

11

Biotechnology Research and Development

In the 1980s the Government of India considered the need for creating a separate institutional framework to strengthen biology and biotechnology research in the country. Scientific agencies supporting research in modern biology included: Council of Scientific and Industrial Research (CSIR), Indian Council of Agricultural Research (ICAR), Indian Council of Medical Research (ICMR), Department of Science and Technology, and University Grants Commission. Biotechnology was given an important boost in 1982 with the establishment of the National Biotechnology Board. Its priorities were human resource development, creation of infrastructure facilities, and supporting research and development (R&D) in specific areas.

The success and impact of the National Biotechnology Board prompted the Government to establish a separate Department of Biotechnology (DBT) in February 1986. There have been major accomplishments in areas of basic research in agriculture, health, environment, human resource development, industry, safety, and ethical issues.

Basic research is essential on all aspects of modern biology including development of the tools to identify, isolate, and manipulate the individual genes that govern the specific characters in plants, animals, and microorganisms. Recombinant DNA (rDNA) technology is the basis for these new developments. The creativity of the scientists and the basic curiosity-driven research will be the keys to future success. India led through the work of G.N. Ramachandran, in which he elucidated the triple helical structure of collagen. The Ramachandran plot has proven to be fundamental in solving the protein structure.

Areas of biosystematics using molecular approaches, mathematical modeling, and genetics including genome sequencing for human beings, animals, and plants, will continue to have priority as we move into the next century. The tremendous impact of genome sequencing is increasingly evident in many fields. As an increasing number of new genes are discovered, short, unique, expressed sequenced tags segments are used as signatures for gene identification. The power of high throughput sequencing, together with rapidly accumulating sequenced data, are opening new avenues in biosciences.

In the plant genome area, the sequencing of Arabidopsis and rice genome will soon be completed and cataloging and mapping of all the genes will be done. There have been major achievements in basic bioscience in the last decade or so in India, where we have expertise in practically all areas of mod-ern biology. The institutions under the CSIR, ICMR, ICAR, DST, and DBT have established a large number of facilities where most advanced research work in biosciences is being done.

In the identification of new genes, development of new drug delivery systems, diagnostics, recombinant vaccines, computational biology, and many other related areas, considerable success has been achieved. Breakthroughs include studies on the three-dimensional structure of a novel amino acid, a long protein of mosquito

(University of Poona), and demonstration of the potential of the reconstituted Sendai viral envelops containing only the F protein of the virus, as an efficient and site-specific vehicle for the delivery of reporter genes into hepatocytes (Delhi University).

The post Green Revolution era is almost merging with the gene revolution for improving crop productivity and quality. The exploitation of heterosis vigor and development of new hybrids including apomixis, genes for abiotic and biotic resistance, and developing planting material with desirable traits and genetic enhancement of all important crops will dominate the research agenda in the next century.

Integrated nutrient management and development of new biofertilisers and biopesticides would be important from the view-point of sustainable agriculture, soil fertility, and a clean environment. Stress biology, marker-assisted breeding programmes, and studying the important genes will continue as priorities.

In India we have achieved the cloning and sequencing of at least six genes, developed regeneration protocols for citrus, coffee, mangrove species, and new types of biofertiliser and biopesticide formulations, including mycorrhisal fertilisers. Research to develop new genetically improved (transgenic) plants for brassicas, mung bean, cotton, and potato is well advanced. Industries have also shown a keen interest in the options of biotechnology and are participating in field trials and pilot level productions.

The successful tissue culture pilot plants in the country, one at TERI in New Delhi and the other at NCL in Pune are now functioning as Micropropagation Technology Parks. This has given a new direction to the plant tissue culture industry. The micropropagation parks serve as a platform for effective transfer of technology to entrepreneurs, including training and the demonstration of technology for mass multiplication of horticulture and trees. Considerable progress has been made with cardamom and vanilla, both important crops. Yield of cardamom has increased 40 percent using tissue-cultured plants.

Between 1996 and 1998, in just eight countries, the area covered by rew genetically improved transgenic plants (from 16.8 to 27.8 million hectares). Some of the main crops grown are soybean, corn, canola, cotton, and potato. The United States, Argentina, Brazil, and China have moved ahead quickly. The new plants exhibited herbicide, insect, and viral resistance, and overall improvement in product quality.

While the Green Revolution gave us self-reliance in food, the livestock population has provided a "White Revolution," with 80 percent of the milk in India coming from small and marginal farms. This has had a major social impact. A diverse infrastructure has been established to help farmers in the application of embryo transfer technology. The world's first IVF buffalo calf (PRATHAM) was born through embryo transfer technology at the National Dairy Research Institute, Karnal. Multiple ovulation and embryo transfer, in vitro embryo production, embryo sexing, vaccines and diagnostic kits for animal health have also been developed. Waste recycling technologies that are cost effective and environmentally safe, are being generated. The animal science area is also opening up many avenues for employment generation.

With a coastline of more than 8,000 kilometers, and two island territories of Andaman and Nicobar and Lakshadweep, there is great potential for marine resource development and aquaculture. To achieve an annual target production of 10 million metric tons of fish, scientific aquaculture offers great possibilities. In fact, aquaculture products are among the fastest moving commodities in the world. We have to continuously improve seed production, feed, health products, cryopreservation, genetic studies, and related environmental factors. This is an area which will help substantially in the diversification of the breadbasket, and in combating nutritional deficiency.

Food Security

Food security is another area in which biotechnology offers major inputs for healthier and more nutritious food. Millions of people are malnourished, and Vitamin A deficiency affects 40 million children. There are also serious deficiencies of iodine, iron, and other nutrients. A recent UNICEF report on food and nutrition deficiencies in children describes this as a "silent, invisible emergency with no outward sign of a problem." Every year over 6 million children under the age of 5 die worldwide. About 2.7 million of these children die in India. More than half of these deaths result from inadequate nutrition.

With the advent of gene transfer technology and its use in crops to achieve higher productivity and better quality, including improved nutrition and storage properties. To ensure adaptation of plants to specific environmental conditions, to increase plant tolerance to stress conditions, to increase pest and disease resistance, and to achieve higher prices in the marketplace. Genetically improved foods will have to be developed under adequate regulatory processes, with full public understanding.

It is scientifically well established that an environmentally benign way of ensuring food security is through bioengineering of crops. For the 4.6 billion people in developing countries, one billion do not get enough to eat and live in poverty. Is there any other strategy or alternative? Biotechnology will provide the new tools to breeders to enhance plant capacity. Since the 12 percent of the world land is under agricultural crops, it is projected that the per capita availability may be reduced from 2.06 hectares to 0.15 hectare by 2050.

Plant Biotechnology

With more than 47,000 species of plants and two hot-spots of biodiversity, 8 percent of the total biodiversity of the earth is available in the Indian subcontinent. The bioresource and biodiversity constitute the mainstay of the economy of the poor people, and special emphasis is required for plant biotechnology research. Isolation of genes for abundant proteins, combining molecular genetics and chromosome maps, and a much better understanding of the evolutionary relationship of the members of the plant kingdom, have led to the potential of plant species being the major source of food, feed, fiber, medicine, and industrial raw material.

Molecular fingerprinting and areas of genomics and proteonics will penetrate the barriers of fertilisation to allow transfer of important characters from one plant to another. By identifying appropriate determinants of male sterility, we can extend the benefit of hybrid seeds to more crops.

A number of centers for plant molecular biology in different parts of the country were initially responsible for training significant numbers in crop biotechnology. There are innumerable possibilities of producing more proteins, vitamins, pharmaceuticals, coloring material, bioreactors, production of edible vaccines, therapeutic antibodies and drugs. Promising leads are available in these areas, and a number of genetically improved crops are ready for field trials of transgenic plants. Work on developing transgenic cotton, brassica, mung bean, and potato has significantly advanced.

Environmental Protection

A special area of global concern amongst the scientific community is environmental protection and conservation, and the need for a policy of sustainable development in harmony with the environment. The Stockholm Conference in 1972, and the UNCED Conference in Rio de Janeiro in 1992, both focused world attention on areas of pollution, biodiversity conservation, and sustainable development. Plants and microbes are becoming important factors in pollution control. World Bank estimates show that pollution in India is costing almost US$80 billion, as well as the human cost in terms of sickness and death.

New developments such as bioindicators, phytoremediation methods, bioleaching, development of biosensors, and identification and isolation of microbial consortia are priority research areas. Significant work has been done in India, but developing a more biologically oriented approach towards pollution control would be extremely important. Cleaning up the large river systems and ensuring the destruction of pesticide residue in large slums in the city are priorities in which a biotechnological approach would be environmentally safe.

Phytoremediation to remove the high levels of explosives found in the soil has become a reality. Although it was known that some microbes can denitrify the nitrate explosives in the laboratory, they could not thrive on site. this degradative ability from the microbe to tobacco plants, and these have produced a microbial enzyme capable of removing the nitrates.

Biodiversity

The global biosphere can survive only if resource

utilisation is about 1 percent and not 10 percent. The global environment is regulated by climate changes and biosphere dynamics. Knowledge about biodiversity accumulated in the last 250 years is being used by scientists throughout the world. There are many gene banks, botanical gardens, and herbaria for conservation purposes. There are also molecular approaches including DNA fingerprinting for plant conservation. The totality of gene species and ecosystems has become exceedingly important, not only to understand the global environment but also from the viewpoint of the enormous commercial significance of the biodiversity.

Biotechnology is becoming a major tool in conservation biology. Twelve percent of the vascular plants are threatened with extinction. Over 5,000 animal species are threatened worldwide, including 563 Indian species. India also has about 2000 species of vascular plants that are threatened. Biodiversity is under threat, and understanding the scale of this destruction and extinction is essential.

More research is needed on forests, marine resources, bioremediation methods, restoration ecology, and large-scale tree plantations. The last has reached 180 million hectares and may increase substantially in the next decade. Marine resources provide many goods and benefits including bioactive materials, drugs, and food items and must be characterised and conserved.

Medical Biotechnology

A major responsibility of biotechnologists in the 21st century will be to develop low-cost, affordable, efficient, and easily accessed health care systems. Advances in molecular biology, immunology, reproductive medicine, genetics, and genetic engineering have revolutionised our understanding of health and diseases and may lead to an era of predictive medicine. Genetic engineering promises to treat a number of monogenetic disorders, and unravel the mystery of polygenetic disorders, with the help of research on genetically improved animals.

Globally, there are about 35-40 biotechnology-derived therapeutics and vaccines in use and more than 500 drugs and vaccines in different stages of clinical trials. Every year about 12 million people die of infectious diseases. The main killers according to WHO are acute respiratory infection, diarrheal diseases, tuberculosis, malaria, hepatitis, and HIV-AIDS. There are vaccines being developed for many diseases, and diagnostic kits for HIV, pregnancy detection, and hepatitis are being developed. The technologies have been transferred to industry.

The Department of Biotechnology has developed guidelines for clinical trials for recombinant products, which have now been accepted by the Health Ministry and circulated widely to industry. Promising leads now exist to develop vaccines for rabies, Mycobacterium tuberculosis, cholera, JEV, and other diseases. Recombinant hepatitis B vaccine and LEPROVAC are already on the market. There is a Jai Vigyan technology mission on the development of vaccines and diagnostics. A National Brain Research Centre is being established to improve knowledge of the human brain and the brain diseases.

The discovery of new drugs and the development of the drug delivery system are increasingly important. Bioprospecting for important molecules and genes for new drugs has begun as a multi-institutional effort. A recombinant vaccine for BCG and hepatitis is being developed. The age-old system of Ayurveda practiced in India needs to be popularised and made an integral part of health care. The global market for herbal products may be around US$5 trillion by 2050.

Industrial Biotechnology

Advances in biotechnology can be converted into products, processes, and technologies by creating an interdisciplinary team. The pharmaceutical sector has had a major impact in this field, as rare therapeutic molecules in the pure form become available. Diagnostics have expanded, with over 600 biotechnology-based diagnostics (valued at about US$20 billion worldwide) now available in clinical practice. The polymerase chain reaction (PCR)-based diagnostics are the most common. Indian efforts in the diagnostic area have been commendable, and it is expected that sales will rise from about US$235 million to US$470 million in the next century.

The consumption of biotechnology products is expected to increase from US$6.4 billion to about US$13 billion by 2000. Industrial enzymes have emerged as a major vehicle for improving product quality. In India a number of groups are gearing up to produce industrial enzymes such as alpha-amylose, proteases, and lipases, increasing three-fold by the end of the century, which will match or surpass the computer industry in size, importance, and growth. India is now producing 13 antibiotics by fermentation. Capacity exists to produce important vaccines such as DPT, BCG, JEV, cholera, and typhoid. Cell culture vaccines such as MMR and rabies, and hepatitis-B, have also been introduced

Bioinformatics

The coming together of biotechnology and informatics is paying rich dividends. Genome projects, drug design, and molecular taxonomy are all becoming increasingly dependent on information technology. Information on nucleotides and protein sequences is accumulating rapidly. The number of genes characterised from a variety of organisms and the number of evolved protein structures are doubling every two years. DBT has established a national Bioinformatics Network with ten Distributed Information Centres (DICs) and 35 sub-DICs. A Jai Vigyan Mission on establishment of genomic databases has been started, with a number of graphic facilities created throughout the country. This system has helped scientists involved in biotechnology research.

Ethical and Biosafety Issues

The bioethics committee of UNESCO established in 1993 has evolved guidelines for ethical issues associated with the use of modern biotechnology. Biosafety guidelines for genetically improved organisms (GIOs) need to be strictly followed to prevent harm to human health or the environment. A three-tier mechanism of Institutional Biosafety Committees has been instituted in India: the Review Committee on Genetic Manipulation, the Genetic Engineering Approval Committee, and the state level coordination committees. It is important to give a clear explanation of the new biotechnologies to the public to allay their fears. New models of cooperation and partnership have to be established to ensure close linkages among research scientists, extension workers, industry, the farming community, and consumers.

Gene transformation is done worldwide with four broad objectives: (a) to develop products with new characteristics; (b) to develop pest and disease resistance; © to improve nutritional value; and (d) to modify fruit ripening to obtain longer shelf life. Thus the aims and objectives are laudable and the tools are available. The new technology does, however, call for a cautious approach following appropriate biosafety guidelines. About 25,000 field trials of genetically modifed crops have been conducted worldwide. The anticipated benefits are better planting material, savings on inputs, and genes of different variet-ies can be introduced in the gene pool of crop species for their improvement.

The potential risks include weediness, transgene flow to nontarget plants, and the possibility of new viruses developing with wider host range and their effects on unprotected species. For crops such as corn and cotton with single gene introductions, there is very little problem expected. When multiple genes are involved scientists have to be more cautious. The time has arrived for a serious look at ethical and biosafety aspects of biotechnology. Researchers, policymakers, NGOs, progressive farmers, industrialists, government representatives, and all concerned players need to come together and share a platform to address the following issues.

— Environmental safety

— Food and nutrition security

— Social and economic benefits

— Ethical and moral issues

— Regulatory issues.

There are about 50 approved MS, postdoctoral, and MD training programmes in biotechnology in progress or just about to start, in different institutions and universities covering most Indian States. Short-term training programmes, technician training courses, fellowships for students to go abroad, training courses in Indian institutions, popular lecture series, awards, and incentives form an integral part of the human resource development activities in India. A special feature of the programme has been that since 1996 many students after completion of their training course join industries or work in biotechnology-based programmes in institutions and laboratories. National Bioscience Career Development Awards have been instituted. Special awards for women scientists and scholarships to the best students in biology help promote biotechnology in India and give recognition and reward to the scientists.

Biotechnology-based activities to benefit the poor and weaker sections and programmes for women have been launched. A unique feature is the establishment of a Biotechnology Golden Jubilee Park for Women which will encourage a number of women entrepreneurs to take up biotechnology enterprises that benefit women in particular. This will also encourage women biotechnologists to develop relevant technologies.

States are taking a keen interest in developing biotechnology-based activities. The States of Uttar Pradesh, Arunachal Pradesh, Madhya Pradesh, Kerala, West Bengal, Jammu and Kashmir, Haryana, Mizoram, Punjab, Gujarat, Meghalaya, Sikkim and Bihar have already started large-scale demonstration activities and training programmes.

The Indian Government has made substantial investments in biotechnology research. Bringing Indian biotechnology products to market will require the involvement of large and small entrepreneurs and business houses. This will require substantial investments from Indian and overseas investors. The worldwide trend is that large companies are becoming major players in development of biotechnology products, and also in supporting product-related biotechnology research.

In the years ahead, biotechnology R&D should produce a large number of new genetically improved plant varieties in India, including cotton, rice, brassicas, pigeonpea, mung bean, and wheat. Tissue culture regeneration protocols for important species such as mango, saffron, citrus, and neem will lead to major commercial activities. Micropropagation technology will provide high-quality planting materials to farmers. Environment-friendly biocontrol agents and biofertiliser packages will hopefully be made available to farmers in such a way that they can produce these in their own fields. The country should be in a position to fully utilise, on a sustainable basis, medicinal and aromatic plants.

The development through molecular biology of new diagnostic kits and vaccines for major diseases would make the health care system more efficient and cheaper. Genetic counselling clinics, molecular probes, and fingerprinting techniques should all be used to solve the genetic disorders in the population. The establishment of ex situ gene banks to conserve valuable germplasm and diversity, and a large number of repositories, referral centers for animals, plants, and microorganisms should be possible. Detailed genetic readouts of individuals could be available. Information technology and biotechnology together should become a major economic force. It is expected that plants as bioreactors would be able to produce large numbers of proteins of therapeutic value, and many other important items. The recent discovery of the gene for recalcitrant species was a landmark event.

In vitro mass propagation can be carried out on any desired species with nonrandom programming. Certainly the 21st century could witness a major increase in new bioproducts generated through modern biology. To achieve the goal of self-reliance in this field, India will require a strong educational and scientific base, clear public understanding of the value of new biotechnologies, and involvement of society in many of these biological ventures.

India has a large research and educational infrastructure comprising 29 agriculture universities, 204 central and state universities, and more than 500 national laboratories and research institutions. It should therefore be possible to develop capabilities and programmes so that these institutions act as regional hubs for the farming community, community, where they can get direct feedback about new technological interventions. It will be equally important to establish strong partnerships and linkages with industry, from the time a research lead has emerged until the packaging of the technology and commercialisation are achieved.

The future impact of biotechnology on industrial development, but this does not yet apply to the less developed countries that lack this infrastructure and industrial strength. In view of the current power of biotechnology and its even brighter future, there is no question that the less developed countries must now position and strengthen their status in biotechnology.

12

New Wave of Agricultural Biotechnology

Pharm and industrial crops represent a new wave of biotechnology crops different from the herbicide-tolerant and Bt crops that have dominated the scene until now.The promised benefits (new and cheaper drugs) could be important to many consumers. At the same time, these crops pose obvious and important risks of contamination of the food supply and the environment with non-food substances.

These crops have precipitated a complex and important debate, especially in the global marketplace. Because engineered crops may present risks, particularly to the environment, some critics believe that the federal government must strengthen the regulatory system governing agricultural biotechnology products so that their risks and benefits can be evaluated carefully before they come to market.

Some of these applications (particularly the pharm crops) promise benefits, such as low-cost drugs, important to consumers as well as farmers. On the other hand, again unlike the plants engineered for new agronomic traits, pharm crops pose obvious risks. Unless stringently regulated, gene escape will put biologically active compounds in many unwanted places in the environment and food supply. No one wants drugs in their corn flakes.

Pharm and industrial crops are plants genetically engineered to produce medical and industrial products, including human and veterinary drugs and biologics and industrial and research chemicals. Crops intentionally treated with genetically engineered viruses that, in turn, produce an industrial or medical substance in the infected plants are also considered industrial or pharm crops.

In general, the term "pharm crops" refers to plants producing drugs or biologics and "industrial crops" to those producing industrial or research chemicals. Pharm and industrial crops are produced by the same methods used to genetically engineer food crops. Briefly, scientists use recombinant DNA techniques to locate and isolate genes of pharmaceutical or industrial interest.

These "transgenes" are then inserted into a crop plant using one of several methods now standard in the industry.The resulting pharm or industrial plant then produces the protein product encoded by the transgene as if it were one of its own naturally occurring genes. Farmers can grow pharm and industrial crops in the same way they do unaltered crops.

For most pharm and industrial uses, scientists plan to extract the novel proteins (or the compounds produced as a result of the function of the novel proteins) from the industrial or pharm crop and purify them before use. In such cases, the new proteins in the crops may or may not be harmful. In some cases, the novel products will be delivered in active form to people or animals in the edible fruit or other parts of the plant.

Scientists and companies are developing and testing a large number of crops producing pharmaceuticals, biologics, and industrial and research chemicals. Pharmaceuticals (or drugs), which are, generally speaking, chemicals intended to diagnose, treat, or prevent diseases, are being produced in a number of pharm plants. Other proteins called "defensins" are being manufactured in plants with hopes that their antimicrobial characteristics will make them useful replacements for antibiotics.Drugs are also being produced as anticoagulants, blood substitutes, and hormones, and for wound repair and the treatment of anemia, liver cirrhosis, cystic fibrosis, HIV, and hepatitis B and C.

Biologics are complex biological products such as antibodies, vaccines, and blood products used in human and veterinary medicine. In the biologics category, pharm plants are producing antibodies against the bacterium that causes tooth decay and for treatment of non-Hodgkin's lymphoma.

Vaccines against hepatitis B, rabies,HIV, malaria, autoimmune diabetes, and cholera are being produced in tobacco, corn, or potato plants.These vaccines will be purified from the plants before being given to humans, most likely by injection or as a pill. Crop-derived vaccines meant to be presented as food (edible vaccines) have been produced against diseases including, among others, measles, polio, diphtheria, yellow fever, and various types of viral diarrhea.These vaccines are being made in tomato fruit and other plant parts that can be eaten raw.

Plant-produced edible vaccines are also being manufactured for veterinary purposes. Scientists recently conducted clinical trials in which pigs were fed a vaccine produced in corn against transmissible gastroenteritis virus (TGEV), one of the most im-portant diseases in swine. And a vaccine meant to promote resistance to mink enteritis virus, canine parvovirus, and feline panleukopenia virus has been made in blackeyed beans.

Industrial Chemicals

Industrial chemicals are compounds used in the manufacture of products like paper, plastics, personal care items, and laundry detergents. Many industrial chemicals are enzymatic proteins that promote the chemical reactions necessary for a particular manufacturing process. Trypsin, an enzyme traditionally isolated from bovine sources and used in large volumes in the detergent and leather industries, for example, and laccase, another enzyme used in making detergents but also in the manufacturing of fiberboard, are being produced in transgenic corn.

Other useful industrial chemicals are the products of chemical reactions driven by enzymes, rather than enzymes themselves. Introducing particular enzymes into industrial crops can result in production of these types of industrial chemicals. For example, the first commercialized industrial crop incorporated a gene from the California bay tree to change the fatty acid biosynthesis pathways in canola.The added gene dramatically altered canola oil composition such that nearly 40% of the oil was comprised of lauric acid, a key raw material in the manufacture of soap,detergents,and cosmetic products. Conventional canola does not contain lauric acid.

Chemical compounds used in research and diagnostic laboratories are referred to as research chemicals.The protein avidin, utilized for purifying other proteins,was the first research chemical commercialized from an industrial crop in 1997. Beta-glucuronidase, another enzyme used extensively for plant molecular biology research, was also commercialized in 1997.

The same plantproduced chemical may be used for different purposes. For example, the enzyme trypsin, discussed above as an industrial chemical, also has medical and research uses.Thus, corn plants engineered to synthesize trypsin are producing pharmaceuticals as well as industrial and research chemicals.

Producing Drugs Industrial Chemicals

For years, manufacturers have used bacterial fermentation systems or mammalian cell cultures to produce industrial chemicals and drugs. More recently, transgenic mammals have been used as "bioreactors" to make valuable proteins in milk. For several reasons, however, a relatively new sector of the biotechnology industry views plants as preferable for protein production. Companies have several reasons for turning to plants biofactories: cheaper production, more flexible manufacturing, and avoidance of controversies associated with animal systems.

Cheaper Production

The biggest reason for turning to engineered crops is cheaper production. Companies believe production costs will be significantly lower than for bacterial fermentation, cell cultures, or animal bioreactors systems.The biggest factor in reducing costs is the high yields of recombinant proteins attainable in transgenic plants. Production costs for corn systems are estimated to be between \$10 and \$100 per gram for proteins that currently cost as much as \$1,000 per gram. Dollar figures based on large-scale tobacco production vary widely from less than \$10 per gram to \$1000 per gram. If realized, the optimistic projections of these companies would represent substantial savings over current costs.Vaccines, for example, can cost thousands of dollars per gram.

Faster, more flexible manufacturing. Using a subsidized commodity like corn and the environment as a production (although notably not a purification/ processing) facility could cut not only production costs but also capital investment in, and the time it takes to increase, manufacturing infrastructure. It could also allow for extremely rapid scale-up (or scale-down) of a production pipeline in response to market or other factors. New drugs could, theoretically, become available sooner.

Less controversy than animal production systems. Because most scientists believe that plants cannot transmit animal pathogens (that can cause diseases like mad cow), the companies creating pharm and industrial plants promote their products as safer than animal-sourced proteins. (Plants as bioreactors present their own risks of product contamination, however, from mycotoxins, pesticides, herbicides or endogenous plant secondary metabolites like nicotine and glycoalkaloids.) Using plants as bioreactors also avoids any animal welfare and certain ethical concerns associated with cloning animals and using them as bioreactors.

Companies that produce pharmaceutical and industrial proteins in corn kernels claim their system alleviates problems of product storage, shipment and purification that often arise with bacterial and animal models.

Potential Societal Benefits of Pharm and Industrial Crops

Companies have high hopes that their products will offer substantial societal benefits including lower drug prices for consumers, drugs unavailable any other way, new value-added products for farmers, and inexpensive vaccines for the developing world.

Lower drug prices. Expected savings on infrastructure and production costs lead companies producing pharm and industrial crops to predict drug prices 10 to 100 times lower than current prices. The cost of treating a patient with Fabry's disease, currently as much as $400,000 a year, for example, is predicted to drop to approximately $40,000 annually. Similarly, it is claimed that the leaves from only 26 tobacco plants could make enough glucocerebrosidase, currently one of the most expensive drugs in the world, to treat a patient with Gaucher's disease for a whole year.

Drugs unavailable any other way. Cheap production also means that drugs that could not be produced cheaply enough at high volume through conventional methods might become economically viable using genetically engineered crops.

Monoclonal antibodies ("plantibodies") fall into this category. One company's idea for such a product is a monoclonal antibody against bacteria responsible for tooth decay, which could be used as a dental prophylactic.A topical therapeutic for herpes, as well as antibodies for the treatment of many other diseases, is also under development.

New Value-added Agricultural Products

Although much about the arrangements between farmers and chemical and drug manufacturers remains unclear, these crops are being touted as a boon to farmers. Some companies, as they solicit contract growers, are promoting these crops as an economically viable alternative to commodity production of corn or tobacco with claims that their "partners realize improved returns relative to other production systems."

Inexpensive, easily delivered vaccines. Food plants engineered to contain pieces of disease agents can function as orally administered vaccines, avoiding the need for injection and syringes. Currently, tomatoes and other vegetables are under development for that purpose. Although the vaccines would still have to be standardized for dose and delivered to patients one at a time, the developers hope that the lower production costs and the convenience of avoiding refrigeration would make the products attractive to the developing world.

Corn

Corn is by far the most popular pharm and industrial crop. Since the early 90's, the US Department of Agriculture has allowed more than 200 field trials of pharm and industrial crops. In nearly three quarters of these tests, corn has been the crop of choice. Other crops tested include tomato, rice, barley, alfalfa, sugarcane, soybean, potato, lettuce, lupine, tobacco, and rapeseed (canola). Significantly, most of these plants are food crops, many are feed crops, and all of them can be ingested in some way.

Industry projects that the market for plant-produced pharmaceutical and industrial proteins could reach $200 billion by the year 2010.Tens of companies are in the midst of developing literally dozens of pharm and industrial crops in order to capture a share of that potential market. One commercial partnership boasts of "more than 15 high value industrial and pharmaceutical products in development."

Currently, most pharm and industrial crop products are still in the field-testing stage where trials usually start with very small plots, from under one acre to 10 acres. As the crops get closer to commercialization, the size of test plots can increase dramatically. For example, during a governmentsponsored workshop held in Ames, Iowa, in April 2000, one California-based company claimed to have received a "policy statement" from the USDA indicating that it was "good to go" on a "thousand acres of production material."

Once companies receive the go-ahead for commercialization, the acres needed to meet market demand will vary considerably from product to product.

Some may require large plantings. One company estimates that filling the current US need for a single specific blood protein, for example, would require a tobacco pharm crop consisting of "thousands or hundreds of thousands of acres." By contrast, some products, such as therapeutic vaccines and certain research chemicals will likely require only tens of acres of pharm or industrial plants to meet the specific demands of those particular markets.

There are no pharm (pharmaceutical or biologic) products on the market, although several are nearing the end of the development pipeline. At least two research chemicals and one industrial chemical have been marketed and others are expected soon. Avidin, isolated from transgenic corn and used as a research chemical,was one of the first recombinant proteins from a pharm or industrial crop commercialized. It has been available from the Sigma-Aldrich Chemical Company since 1997.

Corn-produced beta-glucuronidase, an enzyme used in research laboratories,was also commercialized in 1997. Laurate canola, engineered to produce highlaurate oil useful in soaps, cosmetics, and other products,was approved for use in 1995 but has not been a commercial success. The company producing trypsin in corn plants expected to have limited quantities of this protein, used for industrial, drug, and research purposes, commercially available "in late 2002 with scale up to meet market demand in 2003."

Several vaccines produced in various pharm plants are currently undergoing pre-market testing. Some are in preclinical trials, many are in early clinical tests,and at least one - the non-Hodgkin's lymphoma therapeutic cancer vaccine - was slated for late clinical trials by mid-2002. Regulatory approval for some of these plant-derived medical products is expected by 2004. In Canada, the anticoagulant, hirudin,was briefly grown commercially in transgenic oilseed rape plants.

As with potential benefits and risks, chances of succeeding in the market place and delivering on promises will vary from product to product. For example, while developers of edible vaccines have made progress, they have encountered many obstacles.Whole fruits cannot be used as the delivery system, for example, because detached plant organs continue to be metabolically active and vaccine levels may therefore change, either increasing or decreasing, with time after harvest. (Potatoes sprouting in storage are a good example of this metabolic activity.)

Processing (at least partially) and batch-monitoring of vaccine-containing fruits would therefore be necessary, making storage and processing-associated costs additional issues that must be addressed before edible vaccines, at least for the developing world, are economically viable. There are also hurdles to overcome in reducing production costs. The most optimistic estimates often do not include research and development, sales-associated, and other costs.

Many, including the president of Baxter Healthcare Corporation's recombinant DNA business, have suggested that purification of some proteins from pharm and industrial plants will pose considerable cost-related challenges. Some analysts have even suggested that extraction and sales of conventional food products like meal, oil, and starch might be necessary to make some pharm and industrial crops economically viable. Purification will be a big issue, especially for the drugs and vaccines.

Standards of purity for vaccines and drugs are very high. Effectively purifying foreign proteins away from plant-produced contaminants and/or agricultural products like pesticides could prove formidable. Even if costs of production are reduced, those savings may not be passed on to consumers as lower prices.Virtually none of the biotechnology food products on the market today in the United States deliver price benefits to consumers.

Risks Associated with Pharm and Industrial Crops

Many of the novel substances produced in pharm and industrial crops exhibit high levels of biological activity and are intended to be used for particular purposes, under very controlled circumstances. None of these substances is intended to be incorporated in food or to be broadcast into the environment.

The magnitude of the risks such crops pose depends on many factors including which chemicals are involved, what organisms or environments are exposed, and the level and duration of the exposure. Humans, animals and the environment at large may be at risk. The major concern at this point is that that nonfood substances will contaminate the food supply. Substances intended for use as human drugs are especially problematic because they are intended to be biologically active in people.

The novel chemicals produced by pharm and industrial plants are generally proteins or short polypeptides. (Some novel proteins, in turn, produce chemicals that usually are not proteins.) Both proteins and non-proteins can harm people and animals. Some proteins can act, among other things, as toxins, hormones, or allergens.

Chemicals vary in form and activity through their production cycle and are not always harmful.To illustrate, hirudin, a powerful blood thinner for treatment of heart attacks and strokes, was grown in canola as an inactive compound: only after extraction and purification from canola seeds did it become biologically active.

In addition, the intended mode of use can affect the likelihood of harm. Just because a substance would be bioactive if injected, for example, does not mean that it would be harmful if ingested. Some of the most problematic substances include toxins, hormones, and allergens.

Toxins. Avidin is an example of a potential toxin. Avidin binds extremely tightly, essentially irreversibly, to biotin, a B vitamin necessary for basic metabolism. The tightness of the avidin-biotin bond makes avidin useful as a tool in research laboratories. Binding biotin effectively inactivates it, which can lead to biotin deficiency in living beings, including humans.

Biotin deficiency in humans affects first the skin, then the central and peripheral nervous systems and intestinal tract. If allowed to progress without treatment, biotin deficiency can be serious. Because biotin is an essential nutrient in almost all organisms - deer, squirrels, mice, and birds -animals browsing in avidin-corn fields may also be affected, as might plants, earthworms, and soil microorganisms.

Hormones. Hormones are of particular concern because they can produce substantial physiological effects in minute doses. Higher-than-normal levels of human growth hormone, for example, can cause giantism in children and acromegaly (overgrowth of bone and connective tissue associated with reduced life expectancy) in adults.

Hormones are often developed as drugs and some are being developed in pharm plants. For example, human growth hormone, used to treat hypopituitary dwarfism in children, has been produced in transgenic tobacco. Some chemicals may mimic hormones and thereby disrupt endocrine function.

Allergens. Substances that can induce an allergic response are called allergens. Proteins that elicit immune responses upon ingestion in food are called food allergens and adding new proteins to food crops via genetic engineering may suddenly render familiar food dangerous to sensitive populations. Most proteins do not act as food allergens in humans, but the relatively few that do can be very dangerous. Proteins can also elicit immune responses if injected.

Plant-derived pharmaceutical and industrial proteins have the potential to cause unintended immune or sensitization responses in people who might unknowingly consume them as well as in patients treated with them. Plants modify the proteins they produce differently than do mammals and other organisms, and these modifications may affect characteristics related to allergenicity.

Routes of exposure to people and livestock - the food and feed system. Since most pharm and industrial plants are also the crops that provide food for people and feed for livestock, a major concern is the contamination of the food and feed supplies. Pharm and industrial plants have two major routes into the food and feed system: seed mixing and pollen flow.

Seed mixing. Physical mixing of seeds can occur during seed production, on the farm as a result of using the same machinery to plant or harvest both food crops and pharm and industrial crops, during seed storage or transport, or at the grain elevator or mill.

Pollen flow. Pollen flow is a more indirect way of introducing new genes into food.Wind-borne pollen from corn, currently the most popular pharm and industrial crop, can travel more than 50 yards, making possible fertilization of nearby corn, intended for food, by pharm or industrial plants.As a result, unsuspecting farmers who did not plant pharm corn could nevertheless wind up with food and feed crops contaminated with drugs, harvest them, and offer them for sale to mills. If seed corn is contaminated with drug genes, farmers could unknowingly plant food and feed corn containing biologically active proteins.

Pollen flow is an even wider route of exposure when plants like canola are used as pharm or industrial crops. Reportedly, canola pollen can spread as far as 15 miles from a parent plant and can not only fertilize other canola plants, but other crop plants (including turnip, Chinese cabbage, rutabaga and fodder rape) as well.

In situations where farmers save harvested seed to plant the next growing season, the contaminating genes and gene products may persist year after year. Routes of exposure to the environment. Novel pharm and industrial substances have many routes to the environment. Since genetic engineers often cannot completely confine expression of new genes to particular tissues, pharm and industrial proteins will probably be produced throughout the plants.As discussed above, most of the pharm and industrial genes will be carried, and sometimes expressed, in pollen.

Grazing wildlife. Even if all the pollen, as well as the seeds and other plant material, associated with every pharm and industrial crop remained within the confines of the fields in which they were planted,wildlife could still be exposed. As anyone who has attempted to keep deer out of a vegetable garden or birds from fruit trees can attest, animal consumption of field crops is essentially unavoidable.

Pollinators, herbivorous insects, and soil inhabitants. Pollinators and herbivorous insects will undoubtedly frequent pharm and industrial crop fields as well. Microbes and other animals present in the soils of pharm and industrial crop fields will also be exposed to the compounds produced in these plants via a process scientists refer to as "rhizosecretion" (the exudation of substances from plant roots or when the agricultural byproducts of these crops are plowed back into the earth to "dispose" of them through composting.

Weedy relatives. Although corn and soybeans do not have wild and weedy relatives in the United States, canola and other important plants do. Canola pollen flow could create a wild mustard weed capable of producing a pharm or industrial compound, and such a weed would likely need little human intervention, as compared to a crop plant, to perpetuate itself. Pharm and industrial canola crops thereby threaten not only human food and animal feed supplies but also the greater environment with exposure to biologically active compounds.

Bioaccumulation. Some biopharmaceuticals may have physiochemical properties that could make them persist in the environment or cause them to accumulate in living organisms, dramatically increasing their potential to contaminate ecosystems and their capacity to eventually make their way into the human and farm animal food chains.

The following steps to ensure that the food supply is completely protected against contamination from pharm crops:

1. USDA and FDA should jointly set zero contamination of the food supply as the goal of the agencies' pharmcrop policy. Exposing consumers to drugs though food crops is an unacceptable risk to human health. In addition, the discovery of drugs in food items would cause momentous and costly disruptions in the food system.To guard against these serious consequences, USDA and FDA should work together to set zero contamination of the food supply as the goal of the agencies' pharm crop policy.

2. USDA and FDA should establish a public scientific advisory committee on pharm crops to consider and advise the agencies on the full range of measures available to meet the goal of zero contamination of the food supply. To determine which measures or combination of measures-can achieve a zerocontamination goal,USDA and FDA should convene a panel of experts.The agencies should charge the committee with defining and evaluating all available measures and approaches, including at a minimum the ones listed below, for their contribution to preventing contamination of the food supply.

Potential measures include:

Zoning. Growing pharm crops in designated areas, perhaps states or counties, where they would not encounter food crops could prevent outcrossing and severely reduce opportunities for physical mixing.

Spatial separation. If pharm plants are grown at a sufficient distance from food crops, viable pollen will not reach food crops.

Temporal separation. Outcrossing can also be prevented by planting pharm and food crops so that they flower at different times. If plants do not flower in the same time period, pollen produced by the pharm plants will not encounter receptive eggs to fertilize in the food crops.

Disallowing food crops. Both physical mixing and outcrossing would be substantially reduced if only nonfood (and nonfeed) crops were used for pharmaceutical production.

Dedicated machinery and infrastructure. Dedicating farm machinery (planters, combines, trucks) and grain-handling infrastructure (elevators, trucks, railroad cars) solely to pharm crops would decrease the likelihood of physical mixing.

Indoor production. Growing pharm crops indoors would eliminate outcrossing and, if the facilities were used solely for pharm crops,would also avoid physical mixing.

Sterile pollen. Requiring that pharm crops produce sterile or no pollen would help reduce outcrossing.

Engineering chloroplasts. Splicing drug genes into chloroplasts would reduce outcrossing substantially because pollen typically contains few, if any, chloroplasts.

Suicide genes. Genetic engineers may be able to devise suicide-gene cassettes that would reduce the viability of pharm crop seeds.Although sterile, such seeds would still carry the drug products of interest and would be a food contamination problem if moved into the food system. Suicide genes would also not inhibit the expression of drug genes if transferred to neighboring crops via pollen.

The USDA/FDA pharm-crop advisory committee should deliberate in public. Its members should be selected for their expertise in relevant scientific disciplines and crop production.The panel should be balanced to include representatives from academia, the food and pharm-crop industries, consumer and environmental organizations, and organic and conventional commodity crop grower groups.The committee's report should be written by its members and made public in a timely fashion.

3. USDA and FDA should use the advisory committee's report to devise regulatory requirements to be imposed on growers, handlers, and transporters of pharm crops.

Once the government has the results of the committee's work, it should evaluate the cost and feasibility of adopting the options or combinations of options that meet the goal of zero contamination of the food supply. Once it has selected the appropriate measures, the government should impose them as mandatory conditions on the field testing and commercial growth of pharm crops.

4. USDA and FDA should impose a moratorium on field tests and commercial production of engineered pharm crops until they have convened the scientific advisory committee and established a regime that the scientific community believes will assure the goal of zero contamination of the food supply.

The pharm-crop industry is already struggling in the wake of previous contamination episodes: the StarLink Bt-corn contamination of food products in 2000 and ProdiGene Company's mixing of pharm corn with soybeans at a grain elevator in 2002. Too much is at stake to allow more such incidents, which remain possible as long as pharm crops are grown without adequate confinement.To avoid a "StarLink with drug genes," USDA and FDA should impose a moratorium on field tests and commercial production of engineered pharmcrops. That delay should last until they have convened the scientific advisory committee and established a regime that the scientific community believes will assure the goal of zero contamination of the food supply.

13

Role of Science in Poverty Eradication

The relevance of biotechnology and, specifically, agricultural biotechnology to development, is now at the forefront of international interest. The perceived promise and perils of biotechnology are under intense public scrutiny. The debate is widespread, complex and, frequently, inconclusive. The global scale of demographic pressures in the new millennium will be unprecedented.

The world's population is expected to exceed 8 billion by 2025, an increase of 2.0 billion in the next 25 years. Much of the increase will occur in developing country cities, where urban populations will more than triple. There will be many more mouths to feed in complex circumstances. Norman Borlaug calculates that "to meet projected food demands, by 2025 the average yield of all cereals must be 80 percent higher than the average yield in 1990." These increases must come primarily from increasing biological yields, not from area expansion and more irrigation, because land and water are becoming increasingly scarce.

Meanwhile, poverty and hunger remain pervasive in our world of plenty, despite the enormous burst of output and productivity, the dazzling changes wrought by science and technology, and the amazing achievements recorded on the social indicators for so many of the people on the planet. In the 47 "least developed" countries of the world, 10 percent of the world's population subsists on less than 0.5 percent of the world's income. Some 40,000 people die from hunger related causes every day. A sixth or more of the human family has been marginalised.

Comprehending and preparing for unprecedented levels of global population is the first part of the challenge. The second part is to ensure that this population has access to food in adequate quantities at adequate prices, everywhere, at all times. The third is to produce this food in a way that does not destroy the natural resources on which we all depend. This is the triple challenge we face, and the CGIAR is concerned with the last two facets that combine to form the challenge of sustainable food security.

The challenge is both technological (requiring the development of new, high-productivity, environmentally sustainable, production systems) and political (requiring policies that do not discriminate against rural areas in general, and agriculture in particular), and will have to be accomplished at a time when attention to agricultural development and rural well-being is diminishing. An essential aspect of the response to this challenge is to harness all instruments of sustainable agricultural growth. Agricultural biotechnology is one such instrument. It has moved to the center of the development debate fairly recently, and that marks it out for particular interest and, indeed, concern.

Second, the world of science has grown and changed beyond most expectations. Today, a revolution is taking place in the biological sciences. It is fueled by the groundbreaking work in modern molecular genetics, the enormous advances in informatics and computing, and the enormous sums being invested in biotechnology research.

It is truly an exhilarating time for the biological sciences; similar to what physics experienced in the glorious 40 years between 1905 and 1945, when all the concepts were changed, from cosmology to quantum physics, from relativity to the structure of the atoms.

We are decoding the very blueprints of life; we are learning to manage the deployment and expression of genes. So, we live in a time unmatched for the opportunities that science provides. We can dream of new scientific breakthroughs and new products that can help humanity as never before: New higher yielding plants that are more environment friendly, new remedies for killer diseases, edible vaccines, single cell proteins to feed cattle and clean wastes, hyper-accumulating plants to take toxins out of the soil, expanding forests and habitats where more species thrive, and so much more. We can dream of a future of sustainable development where humans thrive in harmony with each other and with the environment. These opportunities are a necessary focus of interest among all of us who believe that the full potential of science has yet to be realised in our continuing efforts to fight poverty, end hunger, and protect the environment.

A third reason for the current scrutiny of biotechnology is that in recent years, agrobiotechnology has exploded into a major private sector activity, mainly in the industrial countries, with possibilities of even greater expansion in the future. The global area planted with transgenic crops was 1.7 million hectares in 1996, 11.0 million hectares in 1997, 27.8 million hectares in 1998, and 39.9 million hectares in 1999. The US led the field in 1999, as before, with 28.7 million hectares of transgenic crops representing 72 percent of the global area, followed by Argentina-6.7 million hectares (17 percent), Canada-4.0 million hectares (10 percent), China-0.3 million hectares (0.1 percent), and Australia and South Africa-0.1 million hectares (1 percent). Mexico, Spain, France, Portugal, Romania, and Ukraine (in that order) completed the roster, each with <0.1 million hectares (less than 1 percent).

Some 82 percent of the world's transgenic crops were grown in OECD countries, about the same as in 1998 (84 percent). The seven principal transgenic crops grown in 1998 were (in descending order) soybean, corn/maize, cotton, canola/rapeseed, potato, squash, and papaya. The global market for transgenic crop products has grown rapidly during the period 1995 to 1999. Global sales from transgenic crops were estimated at US$75 million in 1995; sales tripled in 1996 and again in 1997 to reach US$235 million and US$670 million respectively, more than doubled in 1998 to reach US$1.6 billion and increased by more than a third in 1999 to reach an estimated US$2.1 to US$2.3 billion.

Thus, revenues for transgenic crops have increased by approximately thirty fold in the five-year period 1995 to 1999. The global market for transgenic crops is projected to reach approximately US$3 billion in 2000, US$8 billion in 2005, and US$25 billion in 2010. In the last three years alone, corporations commercialising transgenic crops and involved with seeds, agricultural chemicals, and the life sciences have been engaged in more than 25 major acquisitions and alliances valued at US$15 billion, and this consolidation is expected to continue. So the biotechnology revolution is here. But it has so far been very much the preserve of the richer countries, a fact that has distorted the debate on what biotechnology can do for the poor. Moreover, like physics in the first half of this century, developments in biological science today compel us to confront profound ethical and safety issues, complicated by the new issues of proprietary science.

Many scientific studies have concluded that the promise of biotechnology as an instrument of development lies in its capacity to improve the quality and quantity of crops and livestock, swiftly and effectively. The application of biotechnology can create plants that are more resistant to drought and soil acidity and salinisation. These attributes are critical to the development of agriculture in the poorest areas where soils are poorly endowed. They are also vital at a time when water scarcity is expected to be a major deterrent to development and, perhaps, a threat to life on the planet as we know it. Additionally, plant characteristics can be genetically altered for earlier maturity, increased transportability, reduced postharvest losses, and improved nutritional quality. Vaccines against diseases afflicting livestock are already important products of biotechnological research.

Most of the early products of agricultural biotechnology focus on crop protection. In 1998, transgenic crops that are herbicide tolerant covered about 19.8 million hectares. Use of herbicide tolerant varieties greatly facilitates weed control using certain types of herbicide and greatly reduces the amount of herbicide applied to the crop for effective weed control. This also enables farmers to employ soil conservation practices such as minimum tiilage. Decreased tillage reduces soil erosion.

Increased plant resistance to pests has also been a major focus of agricultural biotechnology research. In 1998, an estimated 7.7 million hectares were planted to transgenic crops with introduced genes that produce substances toxic to target insect pests. The use of pesticides has dropped in areas using these crops, a positive impact not only on farm income but also on the environment. More recently, research conducted at the Swiss Federal Institute of Technology's Institute for Plant Science has shown that serious problems of malnourishment can be tackled by genetic engineering. Researchers have been able to modify rice grains genetically, to improve the supply of iron and vitamin A in the human diet.

The genetically improved rices can help to reduce global rates of iron deficiency anemia (IDA) and vitamin A deficiency (VAD), especially in developing countries where the major staple food is rice. IDA and VAD are major contributors to childhood blindness and maternal mortality and morbidity primarily in developing countries. Agricultural biotechnology programs, some of them substantial and some, only exploratory, have been established in Brazil, Burundi, China, Colombia, Costa Rica, Cote d'Ivoire, Egypt, Honduras, India, Indonesia, Jordan, Kenya, Malaysia, Mexico, Philippines, Singapore, South Africa, Thailand and Vietnam, amongst others.

Biotechnology has become a lightning rod for an increasingly impassioned debate, with opposing factions making strong claims of promise and of peril. Opposition has been mounted to the spread of transgenic crops or genetically improved organisms (GIOs) and protest movements have developed across the globe. Opposition to biotechnology and specifically to genetic engineering is derived from several viewpoints. They include fears of high-tech farming destroying the livelihood of smallholders, concerns about artificially created products competing with and destroying the marketability of "natural" products, and the presumption of environmental threat.

Many critics fear that biotechnology is a scientists' obsession which is being exploited to bring quick profits to the few even though it can do great harm to the many. Those who hold such views are profoundly concerned that the increased application of biotechnology will harm not only ourselves but even generations of the future. These concerns are genuine and cannot be ignored.

To the extent that we would transform the genetic makeup of a particular variety of plant through genetic transfer from another variety of the same species, that should not pose much of an ethical problem. In fact, it would simply be an accelerated way of achieving by biotechnological means what could be achieved through conventional breeding programs. This process of acceleration should not pose ethical or safety problems for anyone who does not oppose conventional breeding programs.

In all societies, there is a profound distrust of scientists, or anybody else, assuming the right to change the natural order of things. One can argue, rightly, that by our very presence on this planet we are changing the natural order of things, and that our increasing numbers, ever more powerful technology and insatiable appetites for consumption and pollution are indeed affecting nature, and mostly in negative, and potentially dangerous ways. Global warming and biodiversity loss are but two examples. Yet against that general proposition we must set the welfare of the human species.

For instance, hunter and gatherer societies living "in harmony with nature" should be encouraged to stay as they are, with people forced to live in squalor, want, disease, and premature death. A humane treatment of the people would deal with improved diet, education and health, although the resulting reduction in infant mortality and increases in consumption are likely to put pressure on the natural system.

The question then becomes how to handle that pressure, how to ensure that the patterns of development adopted are sustainable, for surely, even arguing from a human-centric point of view, it does not make sense to undermine the ecosystems on which our long-term survival depends. Viewed thus, the matter becomes a calculus of the potential benefits and potential risks associated with change, including the adoption of new technology.

Ethical Issues of Safety Acquire

Ethical issues of safety acquire a different level of concern in the case of GIOs being released into nature. Is there a risk that we would harm the very ecosystems on which all life depends? What if the results of these scientific efforts produce "super weeds" or "super viruses" with an impact so broad that they are devastating? Anxieties about biosafety led to the adoption of restrictive laws in some countries in the 1970s totally banning research into the possibilities of genetic manipulation. These laws have been rescinded, but fears about biosafety continue to bedevil the debate. Here again the question is one of evaluating the scientific evidence and assessing, to the best of our ability, the likely risks and if these can be managed.

Clearly, it is not possible to exclude certain classes of risks entirely, any more than one would be able to exclude the risk of an asteroid hitting the earth or of being struck by lightning. Yet these risks are considered so remote that one in fact goes through life ignoring them. This is not to say that the potential risks of releasing GIOs into the environment are in the same class of probability as either of these two examples, but that the discussion should not start on the basis that any potential risk, no matter how remote, would automatically veto the potential application of a technology.

Patent Rights

Another broad area of ethical issues involved in biotechnology is that of patenting. One of the ethical questions raised is whether the patenting of life forms is acceptable. There is no direct answer for that, but we must take note that the ownership of animals and plants is recognised, so is the right of owning a particular breed. We could argue that allowing ownership rights to other life forms is a matter of degree. After all the varieties of flowers or livestock are themselves owned and sold, breeding of horses and other show animals is recognised, so what is specifically more offensive in patenting, that is establishing an ownership claim, on a gene or gene sequence, in comparison with asserting ownership on a whole animal or plant or variety thereof?

The difference lies in the idea of owning a "building block of life" rather than the actual living creature itself. The assumption being that the building block in question can then be part of many other living things. Nevertheless, the issue is one that affects a lot of people and we should strive to understand their qualms and to accommodate them. No legislation can function if it does not have the broad based support of the majority of the population, and the views of the minority today could well be the majority tomorrow. But such a transformation is best achieved by education and scientific evidence, not by assertive preemptive action by a vocal minority.

Intellectual Property and Knowledge Management

There is another side to the patenting story that raises another set of ethical issues. These include the progressive monopolisation of knowledge and the increasing marginalisation of most of the world population with a concomitant selectivity in focus of research and applications of the new biotechnology and its benefits, skewing it to the potential markets of the rich and excluding the concerns of the poor.

The issues operate at two levels:

— the privatisation of the scientific research enterprise, and the meaning of proprietary science in the coming century; and

— the proprietary aspects of the biotechnology in terms of both process and product.

Also very much at issue are patenting and intellectual property rights (IPR). Supporters of patenting point out that if the private sector is to mobilise and invest large sums of money in agricultural biotechnology R&D, it has a powerful claim to protecting and recouping what it has put into the exercise. On the other side of the argu- ment is the fear that patenting and the exercise of IPR will lead to a monopolisation of knowledge, restricted access to germplasm, controls over the research process, a selectivity in the focus of research and, thereby, the increasing marginalisation of the majority of the world's population.

Regulatory Arrangements

These concerns are complicated by the fact that safeguards governing agricultural biotechnology are uneven and, in some countries, non-existent. Safeguards involve regulatory mechanisms and this calls for actions by governments to put them in place, and ensure that they function effectively.

In the US, the Food and Drug Administration decided in 1992 that genetically improved products would be subjected to the same scrutiny, and be required to maintain the same standards, as all other foods. Extra scrutiny is undertaken only with the introduction of "something truly new."

The process in Europe is much more complex, involving a process of third party refereeing, a vote by the European Commission, and legislation by member states of the union. Regulatory traditions are advancing in developing countries, but not as extensively or as speedily as necessary, mainly because of a lack of capacity. However, ideas are being cross-fertilised, through the influence of international institutions, such as the Convention on Biological Diversity, and under the umbrella of the CGIAR. Special efforts have been made by CGIAR centers, the Rockefeller Foundation, OECD, UNIDO, and UNEP to build developing country expertise in this critical area.

A Balance Sheet

The substantial transformation of smallholder agriculture in developing countries is key to meeting the complex and demanding challenges of the new millennium. The production side of agriculture is a necessary but not sufficient condition to meet the challenge of hunger. If the production side is inadequate, however, discussion of other policies and practices becomes largely academic. Productivity-increasing technologies have to be ecologically sustainable, economically viable, and socially equitable.

Agricultural biotechnology is not a magic wand that can replace poverty and hunger with a regime of plenty, but all the available evidence suggests that it can be an effective additional weapon on the development front. Agricultural biotechnology holds promise of effectiveness in increasing crop yields, reducing the need for chemical pesticides that degrade the environment; supporting resource-poor farmers by nurturing the adaptability of plants to harsh growing conditions such as drought, salinity, and extreme temperatures; improving health by introducing desirable nutritional characteristics into new varieties.

The potential benefits of biotechnology should not divert our attention from the real concerns about the application of the new science. All that is scientifically possible is not ethically acceptable. But the issues of bioethics and biosafety, and of intellectual property rights will be ceaselessly and inconclusively debated unless all those concerned have a genuine desire to reach accommodation based on practical realities, not on emotion or ideology.

Safeguards and regulatory mechanisms have not been established in many developing countries because they are essentially a new branch of law that requires the accumulation of new expertise. A much stronger and more focused international effort is required to strengthen regulatory know-how and practice. In fact, capacity building across the whole spectrum of agricultural biotechnology activities is necessary.

Concerns have also been expressed by civil society representatives that a new wave of hightechnology farming will destroy the interests of small farmers in developing countries. This could be particularly detrimental to women in Asia where 60 percent of the farmers are women, hence the need for national and regional policies that take account of the rights and interests of all concerned. The private sector is at the forefront of every aspect of the agricultural biotechnology revolution -from R&D, through product creation, sales and sharing, to the development of regulatory mechanisms.

14

Experimentation Techniques in Biotechnology

Field testing of genetically engineered plants has come a long way since the first attempts of 1985-1986. It has come to be seen as a fully integrated part of all the projects that aim at eventual commercial introduction of such plants, and as an important component of many fundamental research projects in plant biology.

Field experiments with genetically engineered plants are not a goal in themselves but a means to obtain information which should allow the performer of the experiment to make some progress in his or her project. It is therefore the nature of the project in the context of which a field release is carried out that will determine the way in which a trial is organised. The variety of projects that include field trials with transgenic plants has grown substantially, and covers very diverse areas of fundamental and applied research in biology and agronomy. The ultimate goals of such projects are also very different.

Some projects aim at the production of new crop varieties incorporating new genes introduced by genetic engineering. These are the most widely publicised: insect, virus or herbicide resistant plants, plants with altered protein or lipid composition, etc. Most of these projects are executed by industrial corporations, sometimes in collaboration with public research institutes. They also include many of the earliest achievements in plant genetic engineering.

Another rapidly growing group of projects involves the creation of transgenic plants, and the testing of them in some environments is only a step on the road to better understanding the processes governing the development of plants and their interaction with the outside world. In short, genetically engineered plants are generated with one of two broad goals in mind: the production of material products, or the production of knowledge.

Field Trials in Plant Biotechnology

Testing in the field is usually part of a more extensive sequence of tests performed on plants. The relative importance of field tests in such a sequence is determined by the nature of the trait which has been introduced, by the plant species under investigation, and by the ultimate goal of the project.

If the goal of the project is to test the phenotypic expression of a gene involved in some aspect of the plant metabolism, most of the work may actually be done in the laboratory, in growth chambers, or in a greenhouse. This often has the advantage of allowing better control over the environmental conditions in which the plant is tested, resulting in more clear-cut experimental results. In such projects, scientists may actually avoid bringing plants outdoors, since it does not help them in their work. Hence, an important part of in vivo experimentation in plant physiology, phytopathology and even ecology takes place in a controlled environment. However, a controlled environment only allows the testing of those interactions between plants and their environment which are allowed to vary.

That is why no serious model experiment can be interpreted satisfactorily without reference to those conditions which are kept constant, and without considering possible other interactions in an open field situation, regardless of the nature of the plants being studied. It explains why, after nearly a century of research in controlled environments, nearly all the research in plant breeding and agronomy is still done in the field. There is simply no other effective way to integrate all the variable components of the environment when testing a trait expressed in a plant line for its genetic or agronomic potential.

The same applies for genetically engineered plants, even for "simple" characteristics such as monogenic resistance to pests, diseases or herbicides. The detailed reasons for the necessity of testing transgenic lines in the field will be further discussed later, with a number of practical examples. Now, the practical implications of the existing regulatory framework for field testing genetically engineered plants, as compared with other plants.

Regulatory Constraints on Field Tests

It is important to be aware of the fact that agronomic or genetic field testing is a matter of statistics, of working with quantitatively changing parameters. A plant is not either disease resistant or susceptible: it is scored on a continuous scale from completely devastated to (sometimes) almost completely unaffected. The amount of damage varies depending on a wide range of factors which at first sight may seem to have little to do with either the plant or the pest attacking it.

Therefore any plant breeding programme which includes selection for characters such as disease resistance requires very extensive testing programmes under a wide range of natural environments, usually over several years.

The same is true of resistances built into the plant by genetic engineering. However, there are major differences in the way experiments with genetically engineered plants are set up, compared with non-engineered plants.

1) Field tests with genetically engineered plants require a permit. Trivial as this may seem, it is a very major new situation for scientists working in agronomy and plant breeding, and it takes time for the research community to adjust. The process of obtaining such a permit is complicated and time-consuming. Moreover, the data required in the application for permits is often quite difficult to find, making the preparation of the application a significant part of the experimental work as a whole.

2) Regulations usually require that experiments with transgenic plants be set up in conditions of isolation from related plants, in order to minimise the chance that the engineered genes "escape" into uncontrolled plant populations. Isolation distances can be quite substantial, especially for species which are considered to be efficient cross pollinators.

3) It is also usually required that all the seeds harvested from the trial should be destroyed (including material from nontransgenic controls), and the site of the trial monitored for several years with particular reference, for example, to the development of offspring from buried volunteer seed.

As a result of these special requirements, it is usually impossible to compare results from experiments with transgenic plants directly with results obtained from more traditional trials. One way to overcome this problem is to include all the non-engineered plant lines in the "transgenic" trials, and to treat the whole trial as a transgenic one. Although this is perfectly feasible for some traits, it creates impossible work loads if undertaken in a conventional breeding trial, where a breeder may well want to compare several tens of lines, only a few of which are engineered.

An additional effect is that it is very difficult to comply with the regulatory requirements for companies or institutions working with engineered lines of their main crop. This apparent paradox is best explained with a practical example. Consider a company specialised in sugarbeet breeding, which has developed transformation technology for its crop. Field trials with the transgenic sugarbeet would be quite difficult to conduct at the main experimental stations of that company, since these stations will have a considerable number of other sugarbeet trials on those sites, and would therefore usually not meet isolation requirements. This is why so many field releases of transgenic plants have been carried out in environments which are somewhat unusual for the crop under investigation.

A final restriction on work with engineered plants versus traditional material is that regulations differ between countries. Procedures differ even more, and criteria for approval or rejection of an application are sometimes contradictory. This poses serious constraints to multi-site agronomic trials undertaken in several countries simultaneously-a very common type of trial in most breeding programmes with major crops in Europe.

Field Tests in Project Designs

From the above, it can be deduced that any project in plant biotechnology which includes the prospect of field trials should be designed from day one with the restrictions on these trials in mind. To explain why this is so, it is again best to consider a number of practical examples, and to define which early choices in the project have an impact on the acceptability of the resulting plants for field introduction.

Vector Construction

The DNA sequence that carries the desired gene into the plant is often a quite complex construct. Apart from the target gene and its promoter and termination sequence (both of which, like the gene itself, can originate from very different donor organisms), the construct usually contains a gene coding for a selectable marker (with its own promoter) operating during the in vitro culture phase of the project. Often a different marker gene is used to follow the transgenic line in the greenhouse and in the field. If transformation is achieved with the use of Agrobacterium tumefaciens (the most common method) all the pieces of DNA mentioned above are embedded in two border sequences, originating from the Agrobacterium genome.

It is therefore quite common that the DNA inserted in the host plant contains sequences from three to seven totally different organisms. If one of these sequences originates from an organism which is classified as a human pathogen, this can lead to delays in the approval of the recombinant plant for field release, or even to refusal of the application. Another aspect of vector construction which can create serious problems is the presence of "junk DNA". This is a rather unfortunate name for non-coding sequences of DNA, usually leftover pieces of sequences flanking one or more of the genes of interest in the donor organism.

Often, scientists do not go to the trouble of eliminating all the non-essential DNA in the early stages of vector construction, partly because these flanking sequences often provide useful sites for linking different genes together. Another type of junk DNA is a total DNA preparation from a commercial source (e.g. calf thymus

DNA) which is used to protect the DNA of interest during direct transformation experiments. Some of this DNA may become inserted in the genome of the host plant. One of the recent evolutions in the application of regulations on field releases has been the insistence on "clean vectors", containing only DNA of which the function is well understood.

The time lag between the start of vector construction and the first field trial of the transgenic plant is usually two or more years. The cost of starting with vectors unacceptable for field trials is therefore very high, in terms of time and effort wasted. It is important that information on conditions for field releases is widely distributed in the research community. Most industrial corporations entering a research area are aware of these restrictions, or have procedures in place whereby their scientists check up, at the start of a project, on the regulatory framework in which they are asked to operate. The same is not true for most academic institutions, or for many bodies funding biotechnology research.

Choice of Host Plant

The selection of the right recipient for the genes to be engineered is crucial in determining the chances of success of later field evaluation of the transgenic plants. Until recently, transformation methods were not sufficiently advanced to allow the scientist a wide choice of host species and/or varieties. Most of the pioneering work in plant genetic engineering was carried out with tobacco, simply because it was the only species for which the actual production of transgenic plants was not a limiting factor in the research projects. Somewhat later, scientists started using certain tomato and potato lines as alternative "model species". When those early projects reached the stage where plants were ready for testing under open field conditions, it became clear that a lot of the work had been carried out with varieties of these plants that were not suitable for field evaluation.

Tobacco as a species actually turned out to be an unfortunate choice, because it was difficult to grow to maturity in the climate of the regions where much early genetic engineering work was done. This forced scientists to make complex arrangements to conduct field trials sometimes thousands of kilometers away from their laboratories.

A problem of a different nature arose with early potato work, much of which was done with a very old variety called Berolina. The only reason for choosing this variety was that the transformation results were quite good. Unfortunately, it turned out to be a variety that was very difficult to transfer from the test tube to the field. Therefore it was impossible to test the performance of the introduced genes (in this case insect resistance) in the field.

Another much used potato variety has none of these problems, but flowers profusely. Since many of the early field experiments were approved on condition that the plants would not be allowed to flower (to avoid spreading of the recombinant genes), trial fields had to be patrolled daily and flower buds removed during the entire flowering period.

Troublesome as these early problems were at the time, they served the useful function of stressing the need to consider very carefully the choice of recipient plant in any new genetic engineering project. Today, it is a priority in most applied projects to establish sufficiently reliable transformation procedures for elite breeding material of the project's target crop. This is often a major factor in hindering research. Research teams have repeatedly found, to their distress, that solving transformation for one particular line of a given species offers no guarantee that this will be automatically applicable to other varieties. Nevertheless, the list of plant species of major scientific and/or economic interest for which transformation is now routinely possible using elite germplasm is growing rapidly (Table 1).

Data Collection on Newly Produced Plants

Transformation experiments, and the plants produced, are constantly monitored for parameters indicating success of the experiments. One transformation experiment will usually generate several (sometimes hundreds) of independent transgenic plants. The main goal of the monitoring process is to find the line(s) that gave the best expression of the desired trait in the absence of any side effects. Many of these observations are also used in the preparation of requests for field releases. This early work is done in growth cabinets or, more usually, in a greenhouse.

Expression of the Target Gene

The primary interest of the scientist is, of course, to see if the transformed plants express the new traits. This can be done biochemically, by detection of the protein encoded by the engineered gene, or by a measure of its activity. If the new gene also gives the plant a recognisable new phenotype (e.g. insect resistance), bioassays will also be performed. In the case of insect resistance these assays will include target and nontarget insect species. It is quite common to find differences in expression levels of a factor hundred between independent transformants. This is due to the effect on expression of the place where the gene is inserted in the plant genome.

Expression of the Marker Genes

The expression of the marker (or reporter) genes can be

very important for later genetic work on the transgenic lines, especially if the target trait of the project is not readily observable (e.g. improved protein composition of seeds). The most widely used marker genes are the resistance genes to the antibiotic Kanamycin and to the herbicides phosphinothricin and sulfonylurea, and the gene coding for the enzyme glucuronidase.

Gene Copy Number

In most cases it is preferable to obtain the desired effect with only a single copy of the new gene, since this greatly facilitates the subsequent genetic work of transferring the trait in other germplasm by conventional breeding techniques. Plants are usually screened for copy number by Southern blot analysis on the original transformants, followed by a segregation study on the first generation progeny. In those cases where no single copy plants with all the desirable traits are found, multiple copy plants are separated into single copy plants by outcrossing.

Table 1: Routinely Engineered Crops

Major field crops:	tobacco	potato
	tomato	rapeseed
	cotton	alfalfa
	sugarbeet	maize
	soybean	melon
Vegetable crops:	cauliflower	lettuce
	carrot	chicory

Plant Morphology

All plants are carefully observed for visible malformations. These can result from mutations induced by the in vitro growth conditions of the plants, or from the fact that the engineered genes have by chance been inserted in a locus of the plant genome which is important for normal plant development. These observations are usually continued into the second greenhouse-grown generation.

Upscaling of Material for Field Testing

This is a critical part of the greenhouse work. Field tests usually require substantial amounts of seed. More importantly, for most crops the quality of the seed produced in the greenhouse is not as uniform as field-produced seed. For some crops, even the best seed batches produced in the greenhouse will perform less than elite field-grown seed in a yield trial. To make the best of the first field trials in a project, it is essential that utmost care is taken to use uniform seed from the onset of the work.

Execution of the Field Trial

Pre-release activities

The work on a field trial starts about one year before anything is planted, with the prospection of possible sites. This includes checking whether the expected isolation requirements can be met. It may also include preliminary experiments with control plants to verify that the site is suitable for study of the target gene (e.g. presence of insect pests), especially if the trial has to be conducted in a place where the crop is not usually cultivated.

Trial approval

Most countries require about three months to process an application file. In practice, and especially if a previously untested gene or crop is proposed, it is better to introduce the file five to six months before the expected field planting date, to allow for additional questions from the authorities.

Execution of the Field Experiment

The central goal of most field experiments is the continued observation of the new traits of the engineered plants. Methods used, and trial layout itself, will be determined by this. In experiments where the main goal is to compare yield, a conventional statistical design will be used. In disease resistance trials it may be necessary to provide for good inoculation conditions for the disease. There is a large body of experience with this type of trial in plant pathology.

One of the specific features of early trials with engineered plants is the emphasis on further screening of different transformed lines for conformity to the non-transformed control. Although most off-type plants will have been eliminated in the greenhouse stage, subtle side effects of the transformation procedure or of the introduced genes can only be visualised in a replicated growth and yield trial in the field, where the full complexity of a variable environment influences the development of the plants, and where plot size and number is sufficient to allow the often very small effects on growth and/or yield parameters to become statistically measurable. Another aspect of the observations in the field is related to the requirements for monitoring put forward by the authorities in the approval.

The requirements depend on the crop and trait under investigation. Monitoring extends to the period after termination of the experiment, and to the surrounding area of the trial.

Termination activities

After harvesting the trial, the plant material in the field has to be disposed of. It should be kept in mind that in many trials the "harvest" consists of observations and measurements in different stages of development, and many trials are terminated well before the plants have grown to maturity or set seed. Depending on the developmental stage of the plants at termination, different effective methods for destruction are used. If the plants are still green and growing actively, by far the most effective destruction method is by spraying them with a systemic herbicide, which ensures that the roots are destroyed along with the above-ground parts. If the crop is mature and dry at harvest, it is often burned, or rotovated into the soil.

Testing insect resistance

The first field experiments done by Plant Genetic Systems Ltd. (PGS) were on tobacco plants which had been engineered to express a protein of bacterial origin (Bacillus thuringiensis) which kills the caterpillars and other larvae of certain major tobacco insect pests. However, the trials had to be done in the USA (North Carolina) because these pests are not significant in Europe, and therefore no meaningful data on field resistance could be obtained in Europe. The most remarkable result of these trials was that insect control was much better in the field than in previous growth chamber tests.

This quite unexpected observation was studied further by scientists of North Carolina State University. They discovered that the insect resistant plants, which had not been treated with chemical insecticides, carried a much larger population of predator insects and parasites of the pest than control plots that were maintained insect free with conventional insecticides.

Testing promoter activity

One of the most important qualities demanded from any genetic system before it can be considered for commercialisation is stability of performance under different field conditions. In genetically engineered crops, two conditions can cause failure to perform: the instability of the gene product (enzyme), or erratic expression of the gene. The latter is caused by the sensitivity of the promoter to changes in environmental conditions. It is possible to obtain some preliminary information on the stability of enzymes by extrapolating results from in vitro tests (although these should be verified in vivo), but the stability of promoter activity can only be tested reliably in vivo, preferably under the highly variable conditions found in the field.

In two separate field tests, on tobacco and alfalfa, it was demonstrated that two promoters, both of which controlled expression of a herbicide resistance gene, and both of which gave adequate resistance in greenhouse tests, behaved very differently under field conditions, one promoter giving good resistance under all conditions, while the other gave much lower levels of resistance in the field than in the greenhouse.

Predicting behavior of genes and promoters

One of the most appealing qualities of genetic engineering is that, in principle, the same gene will work in the same way in any organism, provided that the signals controlling its expression function equally well in different receptor species. This is why much of the early transformation work on monocotyledonous crops (at present mainly maize) is again using the old and very thoroughly studied gene coding for antibiotic resistance and/or herbicide resistance to evaluate the performance of promoters specifically isolated for work in monocotyledonous species, and to find out if the "old" dicotyledonous and non-plant promoters can be relied on in these new crops. The answer to this last question is important: it could mean years of work to find suitable analogous promoters to the trusted workhorses developed in dicotyledonous plants.

Now that greenhouse evaluation has been completed, the planting season will deliver the critical data to finalise the analysis, in the form of a series of maize trials in Europe and the USA.

Multi-Site Yield Component Trials

Yield is notoriously difficult to measure, because it is an aggregate index of all the influences of the genotype, the environmental conditions and the interactions between these two on the development of plant communities. To separate all these influences, and to single out the effects of differences in genotype, requires testing over a range of environments. Few such multi-site trials have been attempted with transgenic plants, partly because of lack of suitable transformed plant varieties, and partly because of the complexity of undertaking replicated trials in different countries, which impose different confinement and/or monitoring requirements on the replicas, thereby defeating the basic feature of the trial.

An example of such a trial was a two-year, multi-site study of rapeseed expressing an engineered seed storage protein. The material was tested in 1989 in Belgium, Sweden and Canada, and in 1990 again in these three countries and in the UK. The interpretation of the results was complicated by the fact that the variety originally chosen for the transformation was genetically heterogeneous, and therefore gave problems with the choice of suitable control lines for performance evaluation. One of the positive results of this project was that in this case it had been possible to obtain reasonably similar conditions for the release in the different countries.

Testing engineered male sterility genes

The use of conventional male sterile carrier plants for field testing of genetically engineered traits has often been proposed as a good method for eliminating the risk of gene spread through pollen. Recently, male sterility has also been obtained in plants by recombinant DNA technology. The evaluation of this material requires even more multi-site testing than yield testing, since the first requirement of such a system is stability under all the environmental conditions where it may eventually be used. Field work was begun with tobacco in 1989, rapeseed in 1990 and chicory in 1991. It is anticipated that the present range of trial countries for rapeseed testing will be further expanded, as trial results accumulate, to include the full range of climatic conditions where the trait may be used in rapeseed breeding.

Field Trials: A Canadian Experience

Canada has a very large agri-food sector. The 67 million hectares of improved farmland in Canada are divided into 280,000 farms, 98% of which are family-operated production units. The majority of Canada's major field crops have been introduced during the past 200 years (wheat, alfalfa, barley, rapeseed, canola, and soybean). Most of Canada's crops have undergone significant genetic modification through classical plant breeding during the past 100 years.

Due in part to Canada's excellent history in variety improvement (yield improvements, quality improvements, and improvements in pest/stress tolerance), producers are not only receptive to genetic modifications but, in many instances, financially support, through producer associations, the research, development and plant breeding efforts associated with genetic improvement of major crop species. The evolution of recombinant DNA technologies since the mid-1970s, and the emergence of plant transformation technologies in the 1980s, are providing opportunities in Canada for crop improvement through a combination of genetic engineering and classical plant breeding. These opportunities are currently being exploited by a number of public institutions as well as the private sector.

Although Canada had significant experience with plant cell and tissue culture, there was relatively little activity in plant genetic engineering prior to the breakthrough by Jeff Schell, Marc van Montagu and Mary dell Chilton in developing and utilising disarmed Agrobacterium tumefaciens in the early 1980s. At about the same time, the National Research Council's Prairie Regional Laboratory at Saskatoon was transformed into the Plant Biotechnology Institute.

Numerous researchers in public institutions and a few private companies initiated transgenic plant research in Canada. Many public sector faculty positions in plant biology or plant science departments were filled with recruitment trained abroad in molecular biology. Many faculty members also used sabbatical leaves to retool in plant genetic engineering technologies during the mid-1980s. Several small private sector plant biotechnology companies emerged during the early and mid-1980s in Canada, recruiting molecular biologists from the USA and Europe.

Canadian plant scientists worked in networks (both domestic and international) to access technologies and transgenic germplasm for continuing development and utilisation. By the mid-1980s, many public and private sector laboratories were handling transgenic plant material, associated transformation vectors, and genes of potential relevance to Canada's field crop sector.

Field Evaluations

By 1987, it was clear that transgenic plants under evaluation in several laboratories in Canada would eventually need evaluation in simulated production environments under field conditions. In that year Canada's federal Ministry of State for Science and Technology commissioned a background report on regulatory policy options for biotechnology. In the same year, Agriculture Canada reviewed international regulations for genetically engineered organisms and drafted a regulatory process for products of plant biotechnology.

Agriculture Canada, through consultation with the federal Department of the Environment and the Department of Health and Welfare, and through participation in a series of regulatory workshops and consultation with a large number of Canadian public and private sector plant geneticists and breeders, developed a preliminary application procedure in early 1988 for controlled field testing of transgenic plants. The first field evaluation of transgenic plants was conducted later that year by Agriculture Canada. Since then, the evaluation of transgenic plants in Canada has grown steadily and dramatically.

Agriculture Canada reviewed more than 100 applications for field evaluations of transgenic plants in 1991, 302 in 1992, 503 in 1993 and 848 during the first half of 1994 (up to 16th May). The University of Guelph initiated field evaluations of transgenic canola in 1989 and transgenic alfalfa in 1990. To date, transgenic canola and alfalfa families that have been field-evaluated include a variety of alien gene constructs involving DNA from other plant species, microbial species, fungal species and the coat protein of cauliflower mosaic virus.

Characteristics associated with these "transgenes" include antibiotic resistance, herbicide tolerance, modified amino acid composition, male sterility, male fertility restoration, and tolerance to abiotic (physical) and biotic stresses. In all cases, field evaluations of transgenic plants have followed application to, and approval from, Agriculture Canada. Reviews of individual applications have taken from two to four months and have required fairly specific information on the nature of the genetic modifications of transgenes, the experimental procedures that will be employed for the field evaluation, a description (physical and biological) of the field trial locations, and a description of post trial procedures to prevent entry of transgenic plant material into the food chain or agricultural production ecosystem.

Local farmers and the media appeared satisfied that the research was necessary, and in the interests of the Canadian agri-food system generally. Although no subsequent public discussions have been held, trialspecific information is released publicly each year. To date, no trial-specific concerns have been directed to the University as a result of these information releases. In spite of the regulatory framework developed by Agriculture Canada, at least one environmental activist group has criticised the government for permitting field evaluation of transgenic plants. Both the government and the research community (public and private sector) have been criticised for secrecy and irresponsibility associated with field evaluations of transgenic crops. Criticism has been of a general rather than specific nature, intended more to elicit public fear or outrage rather than to address specific concerns regarding human health, environmental protection or sustainable agricultural ecosystems.

Monsanto's Experience in Developing Plant Biotechnology Products

The Monsanto Agricultural Group has had years of experience in conducting field trials with a number of crops improved through plant biotechnology. The early experience gained from field testing of insect-protected cotton plants in the USA pointed towards the importance of Monsanto's policy of being open with respect to communication with the public.

Through modern tools of plant breeding and biotechnology, Monsanto has successfully incorporated a gene from Bacillus thuringiensis (B.t.), a naturally occurring soil microorganism, into cotton so that the cotton plant effectively controls the larvae of caterpillar insects. Monsanto researchers worked for many years before they were ready to take this cotton to the field for its first tests in 1991. During the first year of field testing, 20 smallscale tests covering a total of three hectares were conducted. An essential element in gaining regulatory approval and successfully conducting the tests was open communication with many of the audiences that were interested in this work. These audiences included Monsanto employees, governmental regulators, cooperators, members of the community near the test sites, and members of interest groups.

Openness in many aspects of developments in plant biotechnology has been very important in gaining freedom to operate throughout the earliest stages of work. General public awareness of science and new scientific principles in the USA is not well advanced. However, there is a predominant view in the public arena that science is neither intrinsically good or bad, but is dependent on the people who are undertaking scientific research. Given this situation, public confidence can be enhanced when those involved with the work communicate about it. Early discussions with key community leaders, government officials, educators and others have been well received and have proven invaluable.

In the USA, a number of regulatory agencies in addition to the United States Department of Agriculture (USDA) are interested in field studies, depending on the crop and objective of the study. If the new plants contain pesticide traits the Environmental Protection Agency (EPA) will be involved, and varying requirements are necessary depending on the size of the test and the disposition of the crop products. If the crop products will be used for food and/or feed, the Food and Drug Administration (FDA) may be interested.

At the state level, several states have their own requirements which should be considered prior to field testing. Monsanto provided information to all of the appropriate regulatory agencies very early on in the development of policies and in the conduct of specific studies. One of the consequences of early and regular open communication with these organisations has been the establishment of mutually respectful relationships.

Monsanto has found particular value in working closely with its partners and cooperators in the early planning stages of its field trials. Site selection is important as it involves both agronomic criteria and non-agronomic criteria such as convenience to facilities and security. In plant biotechnology research, the design of this year's trials often depends on the previous year's test results. Unfortunately, the time between the collection and analysis of last year's data and the regulatory deadline for the submission of the paperwork requesting the next round of tests is very short. We have a very narrow time-frame in which to meet the research, planning and administrative requirements. Again, good communication is a key to successful completion of these requirements.

Following test approval, we have a second very narrow time-frame in which to communicate with all of the individuals and organisations that require specific information about the test. We must ensure that all test requirements are met, from shipping, receiving and storing seed to planting conditions.

Biotechnology and Pioneer Hi-Bred

Pioneer Hi-Bred International Inc. is one of the world's largest independent seed companies. Its goal is to deliver improved genetics to the world's farmers. To achieve this goal, the company has strategically placed research stations around the world, including several in South America. Pioneer Hi-Bred's major products are hybrid maize, sorghum, soybean, sunflower, canola, alfalfa and wheat. We expect biotechnology to play a major role in the continued improvement of all these crops, through the technologies of gene mapping and genetic engineering. Pioneer does not see biotechnology as a unique technology producing a unique stream of products, but rather as another tool to be used by its plant breeders in their continuing search for improved germplasm.

Biotechnology in Plant Breeding

In order to be useful to the plant breeder, germplasm derived through biotechnology must be readily available for use in breeding programmes. The scale of breeding programmes for Pioneer's major crops dictates that plant breeders make selections in the field, not in the greenhouse, so the ability to plant genetically engineered material in the field is vital to future crop improvement programmes.

The development of new crop varieties is a finely tuned process. Hundreds of thousands of new genetic combinations are tested each year in the USA alone. Maximum use is made of breeding nurseries in locations such as Hawaii, Puerto Rico and Mexico, so that more than one generation of a crop may be grown in a single year. Potential products are subjected to wide-scale performance testing in areas where they may be sold, including thousands of side by side comparisons in growers' fields. Finally, there is a rapid scale-up to commercial seed production of those few lines that are finally selected for commercialisation.

If genetically engineered materials cannot be integrated into this system, but are subject to unique barriers, whether raised by technical problems, unnecessarily restrictive regulations, bureaucratic delays or unreasonable costs, then the use of biotechnology in the seed industry will be curtailed, ultimately depriving farmers, processors and consumers of the benefits of crops that use fewer chemical inputs and have improved agronomic performance.

Pioneer supports sound science-based regulatory oversight of biotechnology research and product development, for the public confidence that such regulations provide and the high standards they promote. Clear delineation of the regulatory review process for such products will promote further development of the technology. However, regulations designed to safeguard the environment and human health must also be flexible in order to facilitate the conduct of safe research activities.

Field Testing of Maize

Pioneer does not feel that biotechnology poses an entirely new set of safety issues. There is a continuum of concern that should be taken into account when establishing a system of regulatory oversight. In some instances, biotechnology derived products will be directly comparable with traditionally derived products. For example, Pioneer has developed herbicide resistant maize hybrids derived from plants carrying a natural mutation in the acetolactate synthase gene. An equivalent product can be derived through genetically engineering the same mutation into the acetolactate synthase gene and transforming the gene back into maize. The genetically engineered version does not seem to raise any unique safety concerns, so why should it be subjected to additional regulatory oversight? Regulatory oversight should be based on the risks, if any, that are posed by the product. Plant breeders have always sought to introduce new characteristics into domesticated plants, often employing wide outcrosses to wild relatives to access those characteristics.

Confidence in the safety of the resulting material is based on the breeders degree of familiarity with the germplasm, focusing particularly on any known toxins associated with the plant, and on the extent to which the genetic contribution of the wild species has been diluted by backcrossing. Typically it might take at least six generations of backcrossing to recover the recurrent parent with the new trait. During this process the genetic contribution of the wild species has been reduced from 50% to less than 1%, representing perhaps 500-1000 genes. By these standards, changes brought about by the genetic engineer, who introduces only two or three genes, are amazingly focused.

Many "new" genes will actually be somewhat familiar, because they are derived from plants with a long history of consumption as food or feed, but which are being moved into different crops. Examples might include the methionine-rich seed storage protein gene from Brazil nuts (Bertholetia excelsa) that Pioneer has introduced into soybeans in order to enhance the levels of essential amino acids; or wheat germ agglutinin, a plant lectin that inhibits the growth of European corn borer, an important pest of maize. Pioneer sees little inherent risk in field testing such genes in these crops, but agrees that it should proceed with caution until questions about the possible adverse effects of altered patterns of consumption of these proteins (e.g. potential food allergies) or impact on nontarget species have been resolved.

The North American Experience

Pioneer has developed the capacity to genetically engineer most of its key product crops. It has field tested transgenic alfalfa, maize, canola, soybean and sunflower in North America. All these trials have been reviewed by the United States Department of Agriculture (USDA) or by Agriculture Canada, as well as officials from the state or province in which the trials were conducted.

The test protocol also has to be approved by the Institutional Biosafety Committee, a group of Pioneer scientists and managers with experience in biotechnology, agriculture and risk management, and two outside experts in related disciplines. Pioneer has also taken steps to inform local opinion leaders and the press of its activities. Undoubtedly, much of the success of the field testing programme in the USA is due to the scheme of regulatory oversight, outlined in the 1986 Coordinated Framework for the Regulation of Biotechnology, as implemented by the Biotechnology, Biologics and Environmental Protection unit of USDA's Animal and Plant Health Inspection Service (APHIS).

By focusing on scientific principals of risk assessment and avoiding judgments based on speculation, they have earned the respect of the regulated community and discouraged the proliferation of local regulations. Local regulations, while often well intentioned, are seldom consistent and frequently duplicative of national regulations. From the applicants standpoint, the advantages of dealing with a single regulatory entity are considerable.

The USDA system is certainly a model worthy of serious consideration by any country seeking to implement regulatory oversight of transgenic crops. However, it is not perfect, and some of the problems will be discussed in the context of Pioneer's experiences in field testing genetically engineered maize.

Field Testing Genetically Engineered Maize

Maize represents the prime target for most major seed companies but has proved a demanding technical challenge. Within the last two years, Pioneer and several other companies have finally achieved stable transformation of maize. This has been possible through use of the particle gun to accelerate microprojectiles, bearing recombinant DNA, into plant cells, coupled with tissue culture responsive cell lines. Initially, regulators expressed some concern as to the stability of recombinant genes introduced using this technology.

This resulted in demands to see molecular and genetic evidence, in the form of Southern blots and segregation data, confirming stable integration. Recombinant DNA introduced in this way has not proved unstable and does not replicate independently without a plant origin of replication. Pioneer does not have the resources to conduct molecular or genetic characterisation of every construct that may be field tested. After all, it may take 100 independently transformed lines to identify one that has adequate expression of the recombinant gene under field conditions.

If molecular and genetic characterisation is required for regulatory purposes, then it should be done on those lines that are finally selected for commercial advancement. The genes that Pioneer has introduced into maize and has (or will shortly) field test include two herbicide resistance marker genes, two marker genes controlling pigmentation in maize, two dominant inhibitors of gene expression both derived from maize genes, three different viral coat protein genes that may confer resistance to important maize viruses, and wheat germ agglutinin.

Pioneer does not consider that any of the genes described represent a high risk, either in terms of enhancing the weedy properties of wild relatives of maize or in terms of human or animal health. However, Pioneer is conscious that some of these genes, and the control sequences used to achieve expression, do not occur in domesticated maize, and care should be taken to limit dissemination of these genes through seed or pollen until their properties have been adequately evaluated.

Maize itself is a highly domesticated crop and there are no feral populations. It would take fundamental changes in the biology of the plant to make it an invasive weed. There are no wild or weedy sexually compatible relatives of maize in North America, so there is no genetic bridge by which recombinant genes could find their way into wild populations.

15

Ethical Challenges of Agricultural Biotechnology

The words morals and ethics are used to mean roughly the same thing, even though they do not. In its modern definition, morals includes norms also with respect to nature. The discipline of ethics, on the other hand, is moral philosophy -that is, describing the subject as well as comparing and critically reflecting different moralities.

Reflecting philosophically on ethics is a fulfilling and spiritually demanding concern. But ethics including the ethics of biotechnology and genetic engineering-must be brought down from the lofty heights of ideas or values and placed into the reality of everyday life. To deal responsibly means always and above all to deal intelligently-to weigh the consequences of our actions or nonactions according to the benefits and the harm they can provoke.

Intelligent action is acting in one's enlightened self-interest and is thus compatible with the selfish tendencies in some of our societies. To assume that altruism and a holistic world-view are predominant human characteristics would be unrealistic. Because the issue is complex, some essentials are follows:

— Ethical challenges in discussion and with semantics

— Ethical challenges of decision processes

— Ethical challenges with regard to solidarity

At the end of the 20th century, we should see people who have different convictions and opin-ions as fellow human beings with a diverging view of the world. A more humanist attitude amongst those with differing views will also affect the language we use: Wherever managers of multinational corporations or bureaucrats at research institutes dismiss calls for caution or precaution as stupidity or old-style communist rhetoric, they lack not only style but also wisdom. Some observers fear that today a small minority of radical environmentalists are manipulating the public politically, in a way that may deny poorer nations access to a technology that could help them produce more and better food.

A kind of bio-McCarthyism is taking place, leading to slandering and vilification of anybody who sees genetic engineering and green biotechnology as anything but a nail in the coffin of modern society. Complex problems have no simple solutions. Assessing the contribution of genetic engineering to fighting hunger in developing countries is not simply an academic task involving facts and figures and rational evaluation. The interpretation of data is subject to the interests and value judgments of a variety of stakeholders. Because we live in a world of heterogeneous social systems, with a multitude of value judgments and pluralism of interest, identical information leads to diverging verdicts.

Whereas some people consider genetic engineering something unnatural and inherently nasty and a threat to development in poor countries, others see a compelling moral imperative to develop genetically improved crops to combat poverty and ensure food security. The notion that there is no such thing as one objective reality but a multitude of subjective realities seems prevalent in discussions of biotechnology, as it does in discussions of all major social issues.

There is an assumption that science is neutral and objective. What objectivity means here is that the scientist should provide disinterested information about facts, and not permit an intrusion of his or her subjective values. Disinterested sciences have never existed, and never will. It is impossible to avoid having personal valuations affect our judgment, so we should at least make them explicit and give transparency to what we define as desirable and undesirable.

The private sector is often accused of having profit interests. That is true if after investing billions of dollars in research, you hope to find something that is attractive enough that clients will buy it for a good price. Is that a reason for blame in democratic and market-oriented societies, providing the pursuit of commercial interests is based on law and enlightened self-interest? What are the interests of those opposing biotechnology?

For many observers, large international nongovernment organisations are selfstyled Robin Hoods interested in saving the world. This may be true, or it may not be so simple. It would certainly be interesting to shed more light on the necessity of opinion marketing for the generation of funds for NGOs. Some of the NGOs are also adept in using their power in media and voter terms. As the media are more likely to take up wild stories about the creation of monsters than stories about slow but steady progress toward better crop varieties for resourcepoor farmers, a certain kind of semantics and argumentation has direct relevance for the acquisition of funds. Although any set of personal values can be legitimate from the perspective of its holders, personal values should not necessarily be imposed on others in the sense of prescriptive ethics. This holds especially true for the competition of anthropocentric and biocentric values.

Differences of values and convictions start at a very fundamental level: There are people who oppose genetic engineering for the fundamental reason that human beings should not do what they perceive as playing God. Others give the biosphere as such specific rights-that is, that species boundaries are not to be violated. The ambivalence of technological progress and the fact that a technological innovation can be used for good as well as for ill is neither new nor confined to genetic engineering and biotechnology. Whether you see biotechnology as a threat or as a blessing depends in part on where you position human beings in the biosphere.

Technology Inherent Risks

For genetically improved organisms, the risks classified as inherent in the technology are frequently summarised as biosafety risks. There is a wealth of scientific literature on the deliberate release of living modified organisms into either new environments or areas where they could prove particularly harmful. Until today, no severe biosafety risks have become known. The same is true for genetically altered food: Thousands of scientific papers have demonstrated the safety of the technology and no scientifically reputable test has produced so far any hint that genetically improved food could be in any way toxic.

There is a broad consensus amongst most scientists that serious concerns about the release of living modified organisms are unwarranted. In 1999, nearly 41 million hectares around the world were planted commercially with new genetically improved crops, and no serious issue arose. It is particularly cynical that field trials that could prove the ongoing validity of the scientific consensus on safety in the environment are being vandalised, thus preventing the accumulation of further evidence of the behavior of the new varieties.

Most countries with biotechnological-based industries have sophisticated legislation in place intended to ensure the safe transfer, handling, use, and disposal of such organisms and their products. But even with the best procedures and regulations in place, some risks will remain. Risks-calculable risks-must be taken, otherwise technological progress becomes impossible. There is always the possibility, no matter how slim, that something could go wrong. But science deals in probabilities, while the public has little appreciation for P values, so the few studies discussing specific risks have received disproportionate media play.

Technology Transcending Risks

Technology-transcending risks, as opposed to technology-inherent risks, emanate from the political and social context in which a technology is used. In developing countries, these risks spring from both the course the global economy takes and country-specific political and social circumstances. The most critical risks have to do with three issues: aggravation of the prosperity gap between industrial and developing countries, growth in the disparity in income and wealth distribution within poor societies, and loss of biodiversity. This is not the place to go into a detailed discussion of these issues. What has to be stressed again, however, is the necessity to disentangle risks.

Where there is war, civil strife, and harsh political regimes, there will be hunger. Food insecurity is one of the most terrible manifestations of human deprivation and is inextricably linked to every other facet of development. Poverty is one of the major causes of food insecurity, and sustainable progress in poverty alleviation is critical to improved access to food. Poverty is linked not only to poor national economic performance but also to a political structure that renders poor people powerless. So policy matters of a general nature, and in particular good governance, are of overriding importance for food security.

Progress toward food security also requires a proper macroeconomic framework, and the elements that have been most important for success on the poverty front are known today (see Pinstrup-Andersen and Cohen, this volume). Technology-transcending risks mostly materialise because a gap opens between human scientific technical ability and human willingness to shoulder moral and political responsibility. Today, the risks most likely to inhibit development lie in the political, economic, and social milieu in which technology is applied.

All technological decisions must be the result of a scientific weighing of arguments and be based on a sober and disinterested benefit-risk analysis in a specific situation and within a wider technological portfolio-that is, they have to be decisions based on the ethic of accountability.

Ethics of Accountability

At the moment there are more than 800 million people-mostly women and children-living with chronic malnutrition. In addition, hundreds of millions of people more face food shortages during some part of the year. World population will grow by at least another 3 billion over the next 50 years, with virtually all of the increase in developing countries. Researchers at the International Food Policy Research Institute (IFPRI) say that food production in developing countries will have to be doubled in the next 50 years if a major food security crisis is to be prevented.

During that same 50 years, water will become increasingly scarce and what is left will be more polluted. Arable land is shrinking and what is left will be less productive. In addition, Earth is getting warmer, and no one knows what this is going to mean for the ability of poor countries to produce sufficient food. There is one last scary development: Over the past decades, cereal yields per hectare have deteriorated by one-third. Many food experts expect that this downward trend will continue and that conventional breeding might not be able to reverse it.

The spectrum of potential benefits from the application of genetic engineering and biotechnology to food crops in developing countries ranges from diagnostic aids, for example to accelerate the finding of plant and animal diseases, to gene mapping, which allows speedier identification of interesting and useful genetic material for every kind of plant usable in agriculture. The main objective of R&D for food security is to find improved seed varieties that enable reliable high yields at the same or lower tillage costs through qualities such as resistance to or tolerance of diseases and pests as well as to stress factors.

Equally important objectives are the transfer of genes with nitrogen-fixing capacity onto grains, and the improvement of food quality by overcoming vitamin or mineral deficiencies. There is a wealth of serious analyses that see a great potential for genetically improved crops to contribute to human well-being, particularly in developing countries. The possibilities of higher yields from new genetically improved crops plus their capability to cope with soil toxicity may also help prevent the farming of ecologically fragile areas, or the clearing of tropical forests for agricultural purposes. As natural biodiversity in such areas is particularly high, tremendous positive effects for biodiversity would result.

New agricultural technologies can only contribute one stone to the complex mosaic of agricultural development. Policies must ensure that a development-friendly environment exists and that technological progress is oriented toward the needs of the poor, particularly smallholders. All serious analyses admit concerns with regard to human health, environmental safety, and intellectual property rights (IPR), but the majority conclude that-with a proper regulatory regimen enforced-benefits are likely to greatly outstrip concerns, so that ethically there should be every effort to realise these benefits.

Continued research on all aspects of genetic engineering and biotechnology is necessary to maximise benefits and minimise risks. Whatever helps to address public concerns and regain public confidence for genetic engineering and biotechnology must be done, because in the end, in pluralistic democratic societies, it is social acceptance that makes success feasible.

Ethical Dilemmas

Ethical dilemmas are predicaments that force us to decide between two or more alternative courses of action, each of which is more or less fraught with guilt. Tragic situations illustrative of this quandary abound-situations involving life-anddeath decisions and, with them, inevitable suffering and grief. Ethical dilemmas, then, are not situations that confront us with a choice between an ethically enjoined or a forbidden course of action, but rather ones where we are offered a choice between two or more undesirable courses of action. Not doing anything or putting up with a problematic situation can also be a choice, though not an ethically admissible one because it sidesteps the real point at issue: having to decide on which is the lesser evil.

Solutions to ethical dilemmas often demand compromises. Many people feel vaguely uncomfortable with this because of the negative connotations attached to the word compromise-as in an uneasy or a shoddy compromise. But qualms bring us no closer to a solution. In pluralistic societies, it is virtually impossible not to enter into compromises. The law defines merely the ethical minimum. How minimal this is can be seen in the manifest inadequacy of the legal framework in many developing countries, for example, where as a result of institutional deficiencies or the paramount presence of political violence, the law is overridden. So even if the law does not expressly compel it, knowing better imposes the obligation to accept responsibility beyond the letter of the law.

Concretely, if a developing country has no biosafety regulation or has one but does not enforce it, it might be legal to introduce genetically improved crops, because it is not forbidden. It cannot be legitimate, however, as it would not happen with the informed consent of the authorities and farmers in the countries concerned. Over and above innumerable examples of the ineffectiveness of laws, there also exists a clear difference between juridical and ethical accountability. Whereas the juridical is contained within precisely defined bounds, a concern for the whole enjoins that ethical responsibility should not be equally confined. In ethical perspective, not everything that is legal is desirable, and not everything that is desirable is a legal obligation.

Human Solidarity

The world has become more economically polarised both between and within countries: The richest 20 percent of the world saw its share of global income rise from 70 to 85 percent, while the share belonging to the poorest 20 dropped from 2.3 to a mere 1.4 percent. The gap in per capita income between industrial and developing countries more than tripled between 1960 and 1995, from US$5,700 to US$16,168. Will the new technologies deepen these unacceptable inequalities or will they help to reduce them? Looking back at the lessons of the Green Revolution, it seems that the rich got richer, but the poor got less poor.

A new analysis points to the employment and hence income effects of the Green Revolution varieties that eventually raised family income. As there are social differences, such as land ownership and access to credit as well as to irrigation, seed varieties with a higher productivity are likely to increase the income of those who have earlier and better access to the modern inputs. This is why green biotechnology also can only yield social results in line with the social conditions of those who use it. This is regrettable from an equity point of view-but it is a good governance issue, not a biological one.

An improvement of today's poverty situation in developing countries requires not only good governance but also more solidarity from the industrial countries with poor people in poor nations. Through appropriate allocation of resources, international development assistance can help civil society in developing countries to do better. In addition, new and more effective technologies are needed along with research that helps develop such technologies in developing countries. One of the most effective ways of furthering agricultural and hence rural development was and will continue to be bringing cutting-edge research to resource-poor farmers.

Genetic Engineering andf Biotechnology

Genetic engineering and biotechnology are cutting-edge technologies, and where they are appropriate, they can be of great benefit to resource-poor farmers. There is, however, a problem. Many concerned citizens worry that more and more biotechnological research is concentrated in the private sector, and that its results are patented and hence may prove to be too sophisticated or expensive for resource-poor farmers. The worry is justified: When research priorities are determined by the financial return on investment, the needs of those who have the purchasing power are likely to have higher priority than the poverty eradication needs of small farmers. For this reason public research must be strengthened, because its fruits can be passed on to small farmers at cost or, via government channels, even free of charge. This cannot be done with the results of research sponsored by private enterprise.

The Consultative Group on International Agricultural Research (CGIAR), with its focus on the needs of the developing countries, has to continue to play a conspicuous role in such an effort-and international financial support for CGIAR therefore ought to remain high. But there must also be more and more intensive cooperation between the private and public sectors. The special knowledge and know-how and the different experience-and patented intellectual property-at the disposal of the private sector, but used only selectively for lucrative markets in industrial countries could be passed on via donated transfers or favorable licensing terms to public research institutes in developing countries. The feasibility of this has already been demonstrated by a number of concrete examples.

As far as the compensation issue for the use of genetic material from developing countries is concerned, solutions are also within reach. Fair arrangements here are not so much a matter of solidarity but justice. Suppose, for example, that a private seeds company discovered a trait in an Ethiopian barley strain that made it resistant to certain plant diseases, and then genetically transferred this

Lacy, Peter G. *Deploying the Full Arsenal: Fighting Hunger with Biotechnology*. Washington DC: The Johns Hopkins University, 2003.

Murray, David. *Seeds of Concern: The Genetic Manipulation of Plants*. Sydney Australia: University of New South Wales, 2003.

Paarlberg, Robert L. *Issues in Science and Technology. Reinvigorating Genetically Modified Crops*. Richardson TX: University of Texas, 2003.

Shelton, A.M., et. al. *Agricultural Biotechnology: Informing the Dialogue*. Geneva NY: Cornell University, 2003.

Taylor, Michael R. and Jody S. Tick. *Post-Market Oversight of Biotech Foods: Is the System Prepared?* Washington DC: Resources for the Future, 2003.

Tegene, Abebayehu, et. al. *The Effects of Information on Consumer Demand for Biotech Foods: Evidence from Experimental Auctions*. Washington DC: U.S. Department of Agriculture, 2003.

Winston, Mark L. *Travels in the Genetically Modified Zone*. Cambridge, MA: Harvard University, 2004.

Index